AF737443

BY HIS MAJESTY'S COMMAND.

FIELD EXERCISE

AND

EVOLUTIONS

OF

THE ARMY,

AS REVISED BY

MAJOR GENERAL SIR HENRY TORRENS,
K.C.B., AND K.T.S.,

ADJUTANT-GENERAL TO THE FORCES.

PRINTED AND SOLD BY WILLIAM CLOWES,
NORTHUMBERLAND-COURT, STRAND, LONDON.

MDCCCXXIV.

GENERAL ORDER.

Horse Guards,
10th March, 1824.

THE KING having been pleased to decide that one uniform system of Field Exercise and Movement shall be established throughout His Army, the Commander-in-Chief has received His Majesty's Commands to direct, that the Rules and Regulations, approved by His Majesty for this important purpose, and now detailed and published herewith, shall be strictly adhered to, without any deviation whatever; and such Orders, hitherto given, as may be found to interfere with, or to counteract their effect and operation, are hereby annulled.

All General Officers, Colonels, and Commanding Officers of Corps, are therefore held strictly responsible for the due and accurate performance of every part of these Regulations; and in order that no deviation may creep into practice, so as to disturb the exact uniformity, which must be attained and preserved in all Movements, it is His Majesty's further Pleasure, that no Formations shall be executed, except such as are here prescribed, without due and competent Authority; and General Officers on the Staff will report at the periodical Inspections, whether these His Majesty's Commands are strictly complied with.

The Commander-in-Chief has received His Majesty's further Commands to direct, that every Officer shall provide himself with a Copy of these Regulations, and Commanding Officers of Corps are to take care that this Order be duly observed.

By Command of His Royal Highness,
The Commander-in-Chief,

H. TORRENS,

Adjutant-General to the Forces.

INTRODUCTION.

THIRTY-THREE years have elapsed, since the late General Sir David Dundas conferred upon the country the essential benefit of introducing a System of Tactics, calculated to combine and unite the Field Movements of the Forces; and the fundamental principles laid down by that distinguished veteran, must ever form the basis of subsequent systems, and ensure to his memory the lasting gratitude of the British Army. Improvements, however, have been suggested, by practical experience during the late eventful war; which, although important and essential in the abstract, were partially adopted, without adherence to any general or fixed principle of formation: The result was a practice at once desultory, and disunited; and the advantages to be derived from these improvements, were liable to become nugatory by the revival of that discordant variety of system, with its consequent evils, which had called forth the labours of Sir David Dundas; and to the remedy of which the science and accuracy of that eminent Officer had been so successfully applied. Under these circumstances it was deemed necessary by the Commander-in-Chief to take advantage of a period of peace for re-establishing uniformity, by prescribing to the Army a regulated system; which, while it rests on the leading principles introduced by Sir David Dundas, shall combine the additional formations, with the increased celerity in the mode of executing all movements, which have been partially practised for the last few years, at the uncertain discretion of Commanding Officers: And for the attainment of this desirable object, the attention of His Royal Highness has been given, to what appeared to Him, best calculated to unite simplicity with utility, and precision with celerity.

The execution of this task having, at the same time, been intrusted to me, as belonging to the situation in which the King has been pleased to place me, I have proceeded in my humble endeavours to fulfil the important duty so assigned to me, under the gracious protection of His Majesty, and the auspices and instruction of His Royal Highness the Commander-in-Chief: Thus encouraged and supported in the prosecution of my labours, the following Revision of the General Regulations for Field Exercise and Movement has been prepared; and having been confirmed by practice—submitted by the Commander-in-Chief to the King—and finally approved by His Majesty—is now promulgated for the use and guidance of the Army.

H. TORRENS.

CONTENTS.

PART I.

WITHOUT ARMS.

OPEN ORDER.

CLOSE ORDER.

WITH ARMS.

PART II.

OF THE COMPANY.

PART III.

GENERAL PRINCIPLES FOR THE MOVEMENTS OF A BATTALION.

PART IV.

LIGHT INFANTRY.

GENERAL PRINCIPLES FOR LIGHT INFANTRY FORMATIONS.

PART V.

GENERAL PRINCIPLES FOR THE MOVEMENTS OF THE BRIGADE OR LINE.

EVOLUTIONS OF THE BRIGADE OR LINE MOVEMENTS FROM LINE.

CONTENTS.

SECOND LINES.

INSPECTION OR REVIEW.

INDEX.

RULES AND REGULATIONS.

PART I.

INSTRUCTION OF THE RECRUIT.

THE several heads of Instruction for Recruits are to be attended to, and followed, in the manner and order here set forth. The Instructors, to whom this duty is intrusted, and who are to be answerable for its execution, must possess an accurate knowledge of the part each has to teach, and evince such a clear, firm, and concise manner of conveying their instructions, as will command from the men a perfect attention to their directions. They must allow for the weak capacity of the Recruit; be patient, not rigorous, where endeavour and good-will are apparent; for quickness is the result of much practice, and ought not at first to be expected.

Recruits must be carried on progressively; they should comprehend one thing before they proceed to another.—In the first circumstances of position, the firelock, fingers, elbows, &c., are to be justly placed by the Instructor; when more advanced, they should not be touched; but from the example shewn, and the directions given, be taught to correct themselves, when admonished. Recruits should not be kept too long at any particular part of their exercise, so as to fatigue or make them uneasy, and marching without arms should be much intermixed with the firelock instruction. Neither fife, nor music, must on any account be used; it being essential to confirm the Recruit by habit alone in that cadence of step which he is afterwards to maintain in his march to the enemy, amidst every variety of noise and circumstance that may tend to derange him.

In the manner hereafter prescribed, each Recruit must be trained singly, and in squad; and until he is perfect in all points of his duty, he is not to join the battalion;—for one awkward man, imperfect in his march, or distorted in his person, will derange his division, and, of course, operate on the battalion and line in a still more injurious manner. Every soldier, on his return from long absence, must be re-drilled before he is permitted to act in the ranks of his company.

WITHOUT ARMS.

(Plate I.)

Open Order.

Open order is taken by each recruit stretching out his right arm and keeping that distance from his right hand man.

S. 1. *Position of the Soldier.*

The equal squareness of the shoulders and body to the front is the first and great principle of the position of a soldier.—The heels must be in a line, and closed.—The knees straight, without stiffness.—The toes a little turned out, so that the feet may form an angle of about 60 degrees.—The arms hanging near the body without stiffness; the elbows close to the side, the hands open to the front; the little fingers touching the seams of the trowsers. Great care must be taken that the arms are not kept back too much. The belly rather drawn in; and the breast advanced, but without constraint; the body upright, but inclining forward, so that the weight of it may principally bear on the fore part of the feet; the head to be erect, and neither turned to the right nor left.

In order to supple the recruit, open his chest, and give freedom to the muscles, he should be exercised in the use of a wooden club; which ought to be about two and a half feet in length, rounded and shaped to the hand, and of a weight in proportion to the strength of the recruit: He should circle this round his head, continuing it in its vertical position, first with

with the right hand and then with the left: A club will then be put in each hand, and he will circle both round his head alternately. He will also be practised at the extended motions laid down for the Sword Exercise.

Too many methods cannot be used to improve the carriage of the recruit, and banish the air of the rustic. But any excess of setting up, which stiffens the person, and tends to throw the body backward instead of forward, is contrary to every true principle of movement, and must therefore be carefully avoided.

N. B. The words on the margin, which are printed in *Italics*, are the words of command to be given by the instructor.

All words of command, and particularly the words *Halt* and *March*, must be given distinctly and loud.

S. 2. *Standing at Ease.*

Stand at Ease. — On the words *Stand at Ease*, the right foot is to be drawn back about six inches, and the greatest part of the weight of the body brought upon it; the left knee a little bent; the hands brought together before the body; the palms being struck smartly together, and that of the right hand, then slipped over the back of the left; but the shoulders to be kept back and square; the head to the front, and the whole attitude without constraint.

Attention. — On the word *Attention*, the hands are to fall smartly upon the outside of the thighs; the right heel to be brought up in a line with the left; and the proper unconstrained position of a soldier immediately resumed.

When

When the recruit falls in for instruction, he is first to be taught to place himself, on the word *Attention,* in the position above described, to remain perfectly silent and to give his whole attention to his commander. Before the word *Attention* is given, and occasionally during the time of drill, the recruit may be allowed to rest by *Standing at Ease,* as above explained.

When standing at ease for any considerable time in cold weather, the men are permitted to move their limbs, but without quitting their ground, so that upon the word *Attention,* no one shall have materially lost his dressing in the line. In this case the *Stand at Ease* is given in the tone of a permission and not of command.

S. 3. Eyes to the Right.

Eyes Right.	On the word *Eyes to the Right,* glance the eyes to the right with the slightest turn possible of the head. At the words *Eyes to the Left,* cast the eyes in like manner to the left. On the words *Eyes to the Front,* the look and head are to be directly to the front, the habitual position of the soldier.
Eyes Left.	
Eyes Front.	

These motions are useful on the wheeling of divisions,—or in closing, or obliquing, to a flank,—or when dressing is ordered after a halt: and particular attention must be paid in the several turnings of the eyes, to prevent the soldier from moving his body, which should be preserved perfectly square to the front;—but in all marches to the front, the recruit is to be taught to keep his eyes steadily fixed as if looking at some object of his own height at 100 yards distance in front, and the eyes are never to be cast down, or thrown to a flank, except under the circumstances above stated. On all other occasions the touch of the recruit alone must be his guide.

S. 4.

S. 4. *The Facings.*

In going through the facings, the left heel never quits the ground; the body must rather incline forward, and the knees be kept straight.

To the Right, face.	1st. Place the hollow of the right foot smartly against the left heel, keeping the shoulders square to the front. 2nd. Raise the toes, and turn to the right on both heels.
To the Left, face.	1st. Place the right heel against the hollow of the left foot, keeping the shoulders square to the front. 2nd. Raise the toes, and turn to the left on both heels.
To the Right about, face.	1st. Place the ball of the right toe against the left heel, keeping the shoulders square to the front. 2nd. Raise the toes, and turn to the right about on both heels. 3rd. Bring the right foot smartly back in a line with the left.
To the Left about, face.	1st. Place the right heel against the ball of the left toe, keeping the shoulders square to the front. 2nd. Raise the toes, and turn to the left about on both heels. 3rd. Bring up the right smartly in a line with the left.

On

Right, or Left, Half Face.

Fig. D.

On the word of command *Right* or *Left Half Face,* each man will make an exact half face (*a*), as directed, by drawing back or advancing the right foot one inch, by which the whole will stand individually in echellon.

Front.

When it is intended to resume the original front, the word of command *Front,* will be given, and the whole will face, as accurately as possible, to their former front.

Right or Left about three-quarters face.

Front.

When it is necessary to perform the diagonal march to the rear, the recruit will receive the word *Right (or left about) three quarters face,* upon which he brings the ball of the right foot (not the ball of the toe) to the left heel, or the right heel to the ball of the left foot, and makes a three quarters face in the given direction. Upon the word *Front,* if he has faced to the right, he fronts to the left; and if he has faced to the left he fronts to the right.

The feet in the first of the above motions are to be slipped back or brought forward without a jerk; the movement being from the hip, so that the body is kept perfectly steady until faced.

The greatest precision must be observed in these facings, for if they are not exactly executed, a body of men, after being properly dressed, will lose their dressing on every small movement of facing

S. 5. *Position in Marching.*

In marching, the soldier must maintain, as much as possible, the position of the body as directed in Sect. 1. He must be well balanced on his limbs. His arms and hands, without stiffness, must be kept steady by his sides, and not suffered to vibrate. He must not be allowed to stoop forward, or to lean back. His body must be kept square to the front, and thrown rather more forward in marching than when halted, that it may accómpany the movement of the leg and thigh, which movement must spring from the haunch. The ham must be stretched, but without stiffening the knee. The toe a little pointed, and kept near the ground, so that the shoe-soles may not be visible to a person in front. The head to be kept well up, straight to the front, and the eyes not suffered to be cast down. The foot, without being drawn back, must be placed flat on the ground.

Balance Step.

The Recruit being placed in the position of the Soldier, as described above, is instructed in the *Balance Step* in the following manner:—

1*st. Without gaining Ground.*

Caution.	Balance step without gaining ground, commencing with the left foot.
Front.	The left foot is brought gently forward with the toe at the proper angle to the left, the foot about three inches from the ground, the left heel in line with the toe of the right foot.

When

Rear. When steady the left foot is brought gently back (without a jerk), the left knee a little bent, the left toe brought close to the right heel. The left foot in this position will not be so flat as to the front, as the toe will be a little depressed.

Front. *Halt.* When steady, the word *Front* will be given as above, and repeated to the *Rear* three or four times; to prevent the recruits being fatigued, the word *Halt* will be given, when the left foot, either advanced, or to the rear, will be brought to the right.

The instructor will afterwards make the recruit balance upon the left foot, advancing and retiring the right in the same manner.

2dly. *Gaining ground by the word Forward.*

Front. On the word *Front*, the left foot is brought smartly to the front as before; the knee straight, the toe turned out a little to the left and remaining about three inches from the ground. In this posture he remains for a few seconds only in the first instance, till practice has steadied him in the position.

Forward. On this word of command, the left foot is brought to the ground, at 30 inches from heel to heel, while the right foot is raised at the same moment, and continues extended to the rear. The body remains upright but inclining forwards, the head erect, and neither turned to the right nor left.

On

Two. — On the word *Two*, the right foot is brought forward in a line with the left, the toe a little turned out, and the sole quite flat, but raised two inches from the ground.

Front. — On the word *Front*, the right foot is brought forward, and so on.

3dly. *In Double Time.*

The balance step in double time is performed in the manner above described in No. 2, but without the word of command for each step, the instructor merely giving the words *Double time, March*. The recruit judges his own time, going through distinctly the balance of each leg, and when the instructor observes that he is steady, the time is gradually increased to the slow step.

In the balance step, the toe is not to be pointed, or any flourish made with the foot, which is to be placed flat on the ground, without shaking the body.

With a view to determine the exact length of pace required from the recruit in the above movements, recourse will be had to the pace stick, to measure and regulate his step according to the time required.

S. 6. *Slow Step.*

The length of each pace, from heel to heel, is 30 inches, and the recruit must be taught to take 75 of these steps in a minute, without tottering, and with perfect steadiness.

The recruit must be carefully trained, and thoroughly instructed in this step, as an essential foundation for arriving at accuracy in the paces of more celerity. This is the slowest step

step at which troops are to move, and will be applied to movements of parade, and occasionally to the march in line of considerable bodies.

S. 7. *The Halt.*

Halt. — On the word *Halt,* let the rear foot be brought upon a line with the advanced one, so as to finish the step which was taken when the command was given.

N.B. The words *Halt, wheel—Halt, front—Halt, drèss*—are each to be considered as one word of command, and no pause made betwixt the parts of their execution.

Three or four recruits will now be formed in one rank at very open files, and instructed as follows.

S. 8. *Stepping out.*

Step out. — The squad marches, as already directed, in slow time. On the word *Step out,* the recruit must be taught to lengthen his step to 33 inches, by leaning forward a little, but without altering the cadence.

This step is necessary, when a temporary exertion in line, and to the front, is required; and is applied both to slow and quick time: and at the word (slow or quick step) the pace of 30 inches must be resumed.

S. 9.

S. 9. *Stepping Short.*

Step Short. — On the word *Step Short,* the foot advancing will finish its pace, and afterwards each recruit will step as far as the ball of his toe, and no farther, until the word *Forward* be given, when the usual pace of 30 inches is to be taken.

Forward.

This step is useful when a momentary retardment of either a battalion in line, or of a division in column, shall be required.

S. 10. *Marking Time.*

Mark Time. — On the words *Mark Time,* the foot then advancing completes its pace, after which the cadence is continued, without gaining any ground, but alternately throwing out the foot, and bringing it back square with the other.

Forward. — At the word *Forward,* the usual pace of 30 inches will be taken.

This step is necessary when a column, division, &c., on the march, has to wait for the coming up of others.

S. 11. *The Side or Closing Step.*

The side or closing step is performed from the halt in quick time, by the following commands:

Right Close—Quick March.

Left Close—Quick March.

In

Right Close, Quick March.

In closing to the right, on the word *Quick March*, eyes are turned to the right, and each man carries his right foot about 10 inches directly to his right (or, if the files are closed, to his neighbour's left foot), and instantly brings up his left foot, till the heel touches his right heel, and proceeds to take the next step in the same manner; the whole with perfect precision of time, shoulders kept square, knees not bent, and in the true line on which the body is formed. At the word *Halt*, the whole halt, turn their eyes to the front, and are perfectly steady. (Vide *S.* 44.)

Halt.

S. 12. *Stepping Back.*

Step Back, March.

Halt.

The *Step Back* is performed in the slow time and length of pace, from the halt. On the command *Step Back,—March*, the recruit must be taught to move straight to the rear, preserving his shoulders square to the front, and his body erect. On the word *Halt*, the foot in front must be brought back square with the other.

A few paces only of the Step Back can be necessary at a time.

S. 13. *Changing the Feet.*

Change Feet.

To change the feet in marching, the advanced foot completes its pace, the ball of the other is brought up quickly to the heel of the advanced one, which instantly makes another step forward, so that the cadence may not be lost.

This

This may be required of an individual, who is stepping with a different foot from the rest of his division; in doing which he will in fact take two successive steps with the same foot.

S. 14. *Oblique Step.*

To the Left oblique, March.

When the recruit has acquired the regular length and cadence of the slow pace, he is to be taught the oblique step. At the words *To the Left oblique—March,*—without altering his personal squareness of position, he will, when he is to step with his left foot, point and carry it forward 19 inches in the diagonal line, to the left, which gives about 13 inches to the side, and about 13 inches to the front. On the word *Two,* he will bring his right foot 30 inches forward, so that the right heel be placed 13 inches directly before the left one. In this position he will pause, and on the word *Two*, continue to march, as before directed, by advancing his left foot 19 inches, pausing at each step till confirmed in his position; it being essentially necessary to take the greatest care that his shoulders be preserved square to the front. From the combination of these two movements, the general obliquity gained will amount to an angle of about 25 degrees. When the recruit is habituated to the lengths and directions of the step, he must be made to continue the march, without pausing, and with firmness; when he has been made perfect in the oblique step in slow time, he must be instructed in quick time on the same principle.

As

As all marching (the side step excepted) invariably begins with the left foot, whether the obliquing commences from the halt, or on the march, the first diagonal step taken is by the leading foot of the side inclined to, when it comes to its turn, after the command is pronounced.

The squareness of the person, and the habitual cadenced step, in consequence, are the great directions of the oblique, as well as of the direct, march.

Each recruit should be separately and carefully instructed in the principles of the foregoing sections of the drill. They form the basis of all military movements.

S. 15. *The Quick Step.*

The cadence of the slow pace having become perfectly habitual to the recruits, they are now to be taught to march a *quick* time, which is 108 steps in a minute, each of 30 inches, making 270 feet in a minute.

Quick March. The command *Quick March*, being given, with a pause between them; the word *Quick*, is to be considered as a caution, and the whole to remain perfectly still and steady; on the word *March*, they step off with the left foot, keeping the body in the same posture, and the shoulders square to the front; the foot to be lifted off the ground, that it may clear any stones or other impediments in the way; and to be thrown forward, and placed firm; the whole of the sole to touch the ground, and not the heel alone; the knees are to be bent a little so as not to occasion fatigue or constraint.—The arms to hang with ease down the outside of the thigh; the head is to be kept to the front, the body well up, and the utmost steadiness to be preserved.

After

After the recruit is perfectly grounded in marching to the front in quick time, all the alterations of step, as above, for slow time, must be practised in the quick time.

This is the pace which will be applied generally to all movements by large as well as small bodies of troops;—and therefore the recruit must be trained and thoroughly instructed in this essential part of his duty.

S. 16. *The Wheeling Step.*

The *Wheeling Step,* or *March,* is 120 steps of 30 inches each, or 300 feet in a minute. The directions already given for the march in quick time relate equally to this step.

This is applied chiefly to the purpose of wheeling, and is the rate at which all bodies accomplish their wheels, the outward file stepping 33 inches, whether the wheel is from line into column, during the march in column, or from column into line —In this time also, should divisions double, and move up, when passing obstacles in line.

S. 17. *The Double March.*

The directions for the March, in the two preceding sections, apply in a great degree to this Step, which is 150 steps in the minute, each of 36 inches, making 450 feet in a minute.

Double March. On the word *Double March,* the whole step off together with the left feet; keeping the heads erect, and the shoulders square to the front; the knees are a little bent, the ball of the foot only need be brought to the ground. The body is more advanced than in the other marches;

marches; the arms hang with ease down the outside of the thigh, as in the quick march. The greatest care must always be taken that the recruit shall step off at and preserve the full pace of 36 inches, which can be done with ease, if the soldier is properly placed in position, as directed in Section 1; and that the weight of the body inclines well forward on the fore part of the feet.

Halt. As directed in Section 7.

The word *March*, given singly, at all times denotes that *slow time* is to be taken; when the *Quick*, or *Double March*, is meant, the words *Quick*, or *Double*, will precede the word *March*.

Three or four recruits in one rank, with intervals of 12 inches between them, should be practised in the different steps, that they may acquire a firmness and independence of movement.

PLUMMETS, which vibrate the required times of march in a minute, are of great utility, and can alone prevent or correct uncertainty of movement; they must be in the possession of, and constantly referred to by, each instructor of a squad. The several lengths of plummets, swinging the times of the different marches in a minute, are as follow:

		In. Hun.
Slow time	75 steps in the minute	24 , 96
Quick time	108	12 , 03
Wheeling time	120	9 , 80
Double march	150	6 , 26

A musket-ball suspended by a string which is not subject to stretch, and on which are marked the different required lengths, will answer the above purpose, may be easily acquired, and

should

should be frequently compared with an accurate standard in the adjutant's or serjeant-major's possession. The length of the plummet is to be measured from the point of suspension to the centre of the ball.

Accurate distances of steps must also be marked out on the ground, along which the soldier should be practised to march, and thereby acquire the just length of pace.

CLOSE ORDER.

Six or eight recruits will now be formed in rank at close files, having a steady well-drilled soldier on their flank to lead, and may be instructed as follows.

S. 18. *Dressing when halted.*

Dress.

Dressing is to be taught equally by the left as by the right. On the word *Dress*, each individual will cast his eyes to the point to which he is ordered to dress, with the smallest turn possible of the head, but preserving the shoulders and body square to their front. The whole person of the man must move as may be necessary, and bending backward or forward is not to be permitted. He must take short quick steps, thereby gradually and exactly to gain his position, and on no account be suffered to attempt it by any sudden or violent alteration, which must infallibly derange whatever is beyond him. The faces of the men, and not their breasts or feet, are the line of dressing. Each man is to be able just to distinguish the lower part of the face of the second man beyond him.

In

In dressing, the eyes of the men are always turned to the officer who gives the word *Dress;* and who is posted at the point by which the body halts; and who from that point corrects his men, on a point at or beyond his opposite flank.

The faults to be avoided, and generally committed by the soldier in dressing, are, passing the line; the head too forward, and body kept back; the shoulders not square; the head turned too much.

With a view to establish more exactly, the principles on which all dressing depends, the following instructions in the drill of recruits will be observed.

By the Right (or left) forward Dress.

The right hand man will be moved up a pace and a quarter (or half), and another soldier, as a second point, four paces to his right, while the left hand man, or any other person, serves as a corresponding point for the instructor, upon the left. The instructor will then give the word, No. 2, by the *Right, forward Dress,* when the second recruit will take a pace to the front with the left foot, and shuffle up into line with the two points on his right, taking up his touch and dressing at the same time; the instructor, standing clear to the right of the two points, when he sees that the recruit is properly dressed, and the touch perfect, gives the word *Eyes front,* that heads may be replaced and remain square to the front.

Eyes front.

By the Right (or left) backward Dress.

When every recruit individually has practised and is perfect in his dressing up, both by right and left forward, he must be taught to dress back by the right and left in the same manner.

 The

The instructor will then cause two or three recruits to dress up and back together, taking care that the touch is always preserved, and afterwards the whole squad together.

No rank, or body, ought ever to be dressed, without the person on its flank appointed to dress it, determining, or at least supposing a line, on which the rank, or body, is to be formed, and for that purpose taking as his object the distant flank man, or a point beyond such flank, or a man thrown out on purpose;—dressing must then be made gradually, and progressively, from the fixed point, towards the flank one; and each man successively, but quickly, must be brought up into the true line, so as to become a new point, from whence the person directing proceeds in the correction of the others; and he himself, when so directing, must take care that his person, or his eyes at least, be in the true line which he is then giving.

S. 19. *File Marching.*

To the — face. — The recruits must first *face*, and then be instructed to cover each other exactly in file, so that the head of the man immediately before may conceal the heads of all the others in his front. The strictest observance of all the rules for marching is particularly necessary in marching by files, which is first to be taught at the *slow time*, and afterwards in *quick time*.

March. — On the word *March*, the whole are immediately to step off together, gaining at the very first step 30 inches, and so continuing each step without increasing the distance betwixt each recruit, every man locking or placing

placing his advanced foot on the ground, before the spot from whence his preceding man had taken up his,—no looking down, nor leaning backward, is to be suffered, on any pretence whatever,—the leader is to be directed to march straight forward to some distant object given him for that purpose, and the recruits made to cover one another during the march, with the most scrupulous exactness,—great attention must be paid to prevent them from marching with their knees bent, which they will be very apt to do at first, from an apprehension of treading upon the heels of those before them.

S. 20. *Wheeling of a single Rank, in Slow Time from the Halt.*

Right Wheel.

March.

At the word, *Right Wheel,* the man on the right of the rank faces to the right; on the word *March,* they step off together, the whole turning their eyes to the left (the wheeling flank) except the man on the left of the rank, who looks inwards: and, during the wheel, becomes a kind of base line for the others to conform too, and maintain the uniformity of front. The outward wheeling man always lengthens his step to 33 inches, the whole observe the same time but each man shortening his step in proportion as he is nearer to the standing flank on which the wheel is made —during the wheel, the whole remain closed to the

Halt, Dress.

Eyes front.

the standing flank; that is, they touch, without incommoding their neighbour; they must not stoop forward, but remain upright; opening out from the standing flank is to be avoided; closing in upon it, during the wheel, is to be resisted. On the word, *Halt, Dress,* each man halts immediately, without pressing forward. The dressing being completed, the squad receive the command *Eyes front.*

When the recruits are able to perform the wheel with accuracy in the *slow time,* they must be practised in *wheeling time.*

Nothing will tend sooner to enable the recruit to acquire the proper length of step, according to his distance from the pivot, than continuing the wheel without halting for several revolutions of the circle, and also giving the word *Halt, Dress,* at instants not expected, and when only a 6th, 8th, or any smaller proportion of the circle is completed.

S. 21. *Wheeling of a single Rank, from the March.*

Halt, Right Wheel.

Halt, Dress, March.

The recruits are first to be taught to perform this wheeling in *slow time,* and afterwards in the *wheeling time;*—the rank, marching to the front in slow time, receives the word of command, *Halt, Right Wheel;* the man on the right of the rank instantly halts and faces to his right: the rest of the rank turning their eyes to the wheeling flank, as directed in the preceding section, immediately change the step together to wheeling time; as soon as the portion of the circle to be wheeled is completed, the words *Halt, Dress,* will be given, and then *March,* on which the whole rank steps off together at the slow time.

S. 22.

S. 22. *Wheeling backwards, a single Rank.*

On the Right, backwards Wheel.
Quick March.
Halt.
Dress.

At the words *On the Right, backwards Wheel,* the man on the right of the rank faces to his left. At the word *Quick March,* the whole step backward in wheeling time, dressing by the outward wheeling man; those nearest the pivot man making their steps extremely small, and those towards the wheeling man increasing them as they are placed nearer to him. The recruit in this wheel must not bend forward, nor be suffered to look down; but, by casting his eyes to the wheel-ing flank, preserve the dressing of the rank. On the word *Halt,* the whole remain perfectly steady, still looking to the wheeling flank till they receive the word *Dress.*

The recruits should be first practised to wheel backwards at the slow step; and at all times it will be necessary to prevent them from hurrying the pace; an error soldiers are very liable to fall into, particularly in wheeling backwards. This wheeling is necessary to preserve the covering of pivot flanks when large bodies wheel from line into column for the purpose of prolonging the alignement.

S. 23. *Changing the Direction by the Wheel of a single Rank on a moveable Pivot.*

Right (or Left) Shoulders forward.

When the rank is marching to the front, and is ordered to change its direction to either flank, it receives the word *Right* (or *Left*) *Shoulders forward;* upon which the outward file of the named flank continues to step out

at

at the full pace, and the wheel is performed (according to the principle explained in Section 20,) upon the inner file of the other flank, which brings the shoulder gradually round, and gaining ground sufficient to circle round the wheeling point (where such is given) marks time, until it receives the word *Forward*: But the wheel on the moveable pivot is always made at the same time at which the body may be moving. The commander gives the word *Forward*, when he sees that the rank has gained the front on which he intends it to move in a perpendicular direction.

Forward.

WITH

WITH ARMS.

S. 24. *Position of the Soldier.*

WHEN the firelock is shouldered, the person of the soldier remains in the position described (Section 1), except that the wrist of the left hand is turned out, the better to embrace the butt; the thumb alone is to appear in front, the four fingers to be under the butt, the left elbow is a little bent inwards, without being separated from the body, or being more backward or forward than the right one. The firelock is placed in the hand, not on the middle of the fingers, and carried in such a manner that it shall not raise, advance, or keep back one shoulder more than the other; the butt must therefore be forward, and as low as can be permitted without constraint; the fore part nearly even with that of the thigh, and the hind part of it pressed by the wrist against the thigh; the piece must be kept steady and firm below the hollow of the shoulder; should the firelock be drawn back, or attempted to be carried high, in that case, one shoulder will be advanced, the other kept back, and the upper part of the body distorted, and not placed square with respect to the limbs.

Each recruit must be separately taught the position of shouldered arms, and not allowed to proceed until he has acquired it.

S. 25. *Different Motions of the Firelock.*

The following motions of the firelock will be taught and practised as here set down, until each recruit is perfect in them,

them; they being necessary for the ease of the soldier in the ourse of exercise.

As mentioned in the Manual Exercise.	Supporting arms. Sloping arms. Carrying arms. Ordering arms. Standing at ease. Attention. Shouldering from the order.

The recruit must be accustomed to *carry* his arms for a considerable time together; it is most essential he should do so, and not be allowed to *support* or *slope* them so often as is practised, under the idea that long *carrying* them is a position of too much constraint.

A company or battalion is never to come to the HALT, or FORM IN LINE, or to DRESS, (which are situations where the greatest accuracy of front is required,) but with *carried* arms.—When marching in column, or moving by threes, or in file, arms may be *sloped.*—It is to be understood, as a general rule, that in the double march, as the men make the first step, they slope arms, without any separate word of command; on being halted, arms are instantly carried in the same manner.

S. 26. *Attention in forming the Squad.*

When the SQUAD or division (consisting of from six to eight files) *falls in,* each man, with carried arms, will take his place in his rank, beginning from the flank to which he is ordered to form; he will dress himself in line by the rule already given; assume the ordered position of a soldier, and stand perfectly still, and steady, until ordered to stand at ease, or that some other command be given him. Attention must

must be paid that the files are correctly closed: that the men in the rear rank cover well, looking their file leaders in the middle of the neck:—That the rear rank has its proper distance of one pace (30 inches) from the front rank, and that both ranks are equally well dressed:—That the men do not turn their heads to the right or left; and that each man has the proper unconstrained attitude of a soldier.

S. 27. *Open Order.*

Rear Rank take Open Order. The recruits being formed in two ranks at close order, on the word *Rear Rank take Open Order*, the flank men on the right and left of the rear rank, step briskly back one pace, face to their right, and stand covered, to mark the ground on which the rear rank is to halt, and dress at open order; every other individual remains ready to move.—On the

March. word *March*, the dressers front, and the rear rank falls back one pace, dressing by the right

Fig. A. the instant it arrives on the ground.

S. 28. *Close Order.*

Rear Rank take Close Order. *March.* On the word *Rear Rank take Close Order*, the whole remain perfectly steady; at the word *March*, the rank closes within one pace, and then halts.

S. 29. *Manual Exercise.*

According to Regulation.

S. 30.

S. 30. *Platoon Exercise.*

According to Regulation

S. 31. *Firings.*

When the recruits have acquired the management of their arms, and are perfect in the motions of the manual and platoon exercises, they will be instructed at closed ranks in firing.

Direct to their front and both ranks kneeling.

Obliquely to the right and left.

By files.

S. 32. *Marching to the Front and Rear.*

Caution.
March.

The squad, or division, is to be particularly well dressed; files correct; arms carried; the rear rank covering exactly, and each individual to have his just attitude and position before the squad is ordered to move. The march will be made by the right or left flank, and a proper trained man will therefore conduct it. The word *Squad*, or *Division*, may be given as a caution; and at the word *March*, each man steps forward a full pace. The recruit must not turn his head or eyes to the flank by which he is marching, as a turning of the shoulders would undoubtedly follow. His elbows must be kept steady, without constraint; if they are opened from his body, the next man must be pressed upon; if they are closed, there arises an improper distance which must be filled up; in either case waving on the march will take place, and must therefore be avoided.

Turning

Halt, front, March. — Turning to the right or left, or about, in march, is not to be at first practised; but the squad is to *Halt, front,* by command, and then *March.*

A squad, or division, great or small, after a movement to the rear, to a flank by threes, or in file, may immediately resume its proper front; instead of the words to halt, and face about, the word *Halt front,* as one command, will be given, when it is instantly to face to its proper front in line. In general there should be no sensible pause between the *Halt front* of any body; and it is after fronting, that the dressing, if necessary, is ordered to take place.

To march straight forward is of the utmost consequence, and he who commands at the drill will take the greatest pains to make his squad perfect in this essential object;—for this purpose he will often place himself behind the flank file by which the squad is to move in marching, and take a point, or object, exactly in front of that file, and another in its rear; he will then command *March,* and, remaining in his place, he will direct the advance of the squad, by keeping the flank file always in a line with these objects. It is also from the rear, that the leaning back of the soldier, and the bringing forward, or falling back of a shoulder, are soonest perceived; faults which if not instantly rectified, will create confusion in a line, where one man, by bringing forward a shoulder, may change the direction of the march, and oblige the wing of a battalion to run, in order to keep dressed.

In short, it is impossible to labour too much at making the soldier move straight forward, keeping always the same front as when he commenced his march. This is effected by moving solely from the haunches, keeping the body steady, the shoulders square, and the head to the front; and it will be attained without difficulty, by a strict attention to the rules given for

marching,

marching, and a careful observance of an equal length of step, and an equal cadence, or time of march.

The recruit must be practised in changing the pace, without halting, from *slow to quick and double,* and from *quick to slow time;* as well as from *quick to double,* and from *double to quick time:* but never from *double to slow time,* without a previous halt.

Right Turn.
Left Turn.

Turning on the march, in order to continue it, is necessary when companies, or their divisions, are moving in file, and that without halting, it is eligible to make them move on in front; or when moving in front, it is proper without halting to make them move on in file.

Right About, turn.
Left About, turn.
Forward.

This movement is applicable to companies, whereby the front is changed on the march without halting. On the word *Turn,* each individual soldier, without changing step, or cadence, comes to the right or left about on his own ground, and in his own person performing the movement in the time prescribed for three distinct paces, then marking time till he receives the word *Forward,* when he resumes the full pace to the front.

As an aid for fixing the true time, or cadence of the march, the plummet must be frequently resorted to; the words *left, right,* may, when necessary, be repeated. Strong taps of the drum, if in just time, and regulated by the plummet, may be given immediately before the word *March,* to imprint the required measure on the mind of the recruit; but they are on no account, or in any situation, to be given during the march.

S. 33. *Open and Close Order, on the March.*

Rear Rank take Open Order. The squad, when moving to the front in slow time, receives the word *Rear Rank take Open Order;* on which the front rank continues its march, without altering the pace, and the rear rank marks the time, and steps off at the second step.

Rear Rank take Close Order. On the word *Rear Rank take Close Order*, the rear rank steps nimbly up to close order, and instantly resumes the pace, at which the front rank has continued to march.

S. 34. *March in File to a Flank.*

The accuracy of the march in file is so essential in all countermarches and all file movements, that the recruit cannot be too much exercised in it.

To the —face, March. After *facing,* and at the word *March,* the whole squad steps off at the same instant, each replacing, or rather overstepping, the foot of the man before him; that is, the right foot of the second man comes within the left foot of the first, and thus of every one, more or less overlapping, according to the closeness or openness of the files, and the length of step. The front rank will march straight along the given line: each soldier of that rank muss look along the necks of those before him, and never to right or left, otherwise a waving of the march will take place, and, of course, the loss

loss and extension of the line and distance, whenever the body returns to its proper front. The men of the rear rank must look to, and regulate themselves by, their leaders of the front rank, and always dress in their file. Although file marching is in general made in quick time, yet it must also be practised and made in slow time. The same position of feet as above, takes place in all marching in front, where the rear rank is closed, and locked up.

S. 35. *Wheeling in File.*

The squad, when marching in file, must be accustomed to wheel its head to either flank; each file following successively, without losing or increasing distance.—On this occasion, each file makes its separate wheel on a pivot moveable in a very small degree, but without altering its time of march, or the eyes of the rear ranks being turned from their front rank.—The front rank men, whether they are pivot men or not, must keep up to their distance, and the wheeling men must take a very extended step, and lose no time in moving on.

The head of a company marching in file, must change direction in the same manner on the moveable pivot, by gradually gaining the new from the old direction, and thereby avoiding the sudden stop that otherwise would take place.

S. 36. *Oblique Marching in Front.*

Right oblique.

When the squad is marching in front, and receives the word *To the right oblique;* each man, the first time he raises the right foot, will, instead of throwing it straight forward, carry it

in

in the diagonal direction, as has been already explained in Sect. 14, taking care not to alter the position of his body, shoulders, or head. The greatest attention is to be paid to the shoulders of every man in the squad, that they remain parallel to the line on which they first were placed, and that the right shoulders do not fall to the rear, which they are very apt to do in obliquing to the right, and which immediately changes the direction of the front.—

Forward.

On the word *Forward,* the incline ceases, and the whole march forward. In obliquing to the left, the same rules are to be observed, with the difference of the left leg going to the left, and attention to keep up the left shoulder.

The same instructions that are given for slow time serve also for quick time.

In obliquing to the right the eyes are invariably to be turned to the right, but the touch must be preserved to the left and *vice versâ,* excepting in the obliquing of a battalion, when the touch must always be to the centre.

S. 37. *Diagonal March.*

Right (or left) half turn

When the squad is marching to the front, and it is desired to take an oblique direction, the word *Right (or left) half turn* is given, and the men move on the diagonal lines upon which they are individually placed in echellon, as described in the half-facings, S. 4. —and when it is intended to move to the original front without halting, the word *Front*

Front turn.

 turn

Fig. D. *turn* is given, when each man will turn his body to the front and move forwards without checking the pace.

When the movement is performed to the left, the reverse of the foregoing instructions will take place.

During the diagonal march the outer flank will be the pivot for the time being; for instance, when a squad or company is moving by the right half turn, the right-hand man must pay particular attention to the length of pace, and to move perpendicular to the line he took up when he made his half turn, as the accuracy of his movement may assist very much in preserving the division in its proper position: The other files must be careful that their right arms do not get beyond the centre of the men's backs who precede them in echellon; and if they keep this position, their right feet will just clear the left of the preceding file.

Note.—It will be desirable that the instruction for the diagonal march should commence in a single rank without arms.

S. 38. *Wheeling forward from the Halt.*

Right Wheel.
Quick March.
Halt.

The directions already given for the wheeling of a single rank (vide Sect. 20.) are to be strictly attended to in this wheel of the squad. —On the word *Right* (*or left*) *Wheel*, the rear rank, if at one pace distance, locks up. At the word *Quick March*, the whole step together in the wheeling time, and the rear rank, during the wheel, inclines so as to cover the proper front rank men. At the word *Halt*, the whole remain perfectly steady.

S. 39.

S. 39. *Wheeling backward.*

The squad must be practised in wheeling backward in the wheeling time. In this wheel, the ranks may preserve the distance of one pace from each other. Great attention should be paid, to prevent the recruits from fixing their eyes on the ground. (Vide Sect. 22.)

S. 40. *Wheeling from the March, on a halted and moveable Pivot.*

The directions for wheeling on a *halted*, and on a *moveable* pivot, have already been given, in Sects. 21 and 23. The squad should now be practised in both, until the recruits are thoroughly confirmed in those movements.

S. 41. *Stepping out,—Stepping short,—Marking Time, —Changing Feet,—The Side Step,—Stepping back.*

The squad must likewise be practised in *stepping out, stepping short, marking time, changing feet, the side step*, and *stepping back*, the instructions for which have been fully detailed in the foregoing sections.

It cannot be too strongly inculcated, that every just movement and manœuvre depends upon the correct *equality* of *march*, established and practised by all the troops of the same army. When this is not attended to, disunion and confusion must follow, on the junction of several battalions, although, when taken separately, each may be well trained: It is in the

original

original instruction of the recruit, and squad, that this great point is to be attained. The *time* and *length* of step, are prescribed: The TIME is infallibly ascertained, by the frequent corrections of the *plummet*, which, when so applied, will soon give to each man the habitual measure so much desired; and the LENGTH of step is acquired by repeated practice of the pace stick. When a squad marches by files, by threes, or by fours, a man should be placed upon the flank of the leading files, to whose step the pace stick can be conveniently applied, to correct the length of step of the whole.

END OF PART FIRST.

PART II.

OF THE COMPANY

(Plate I.)

S. 42. *Formation of the Company.*

THE Recruit being thoroughly grounded in all the preceding parts of the drill, is now to be instructed in the movements of the company, as a more immediate preparation for his joining the battalion; for this purpose from 10 to 20 files are to be assembled, formed, and told off, in the following manner, as a company in the battalion.

Fig. A.

The company FALLS IN at close order, with shouldered fire-locks; the files lightly touching, but without crowding; each man will then occupy a space of about 21 inches.—The commander of the company takes post on the right of the front rank covered by a serjeant in the rear rank.—Two other serjeants will form a third, or supernumerary rank, (*ss*,) three paces from the rear rank.

When a company is thus singly formed with its officers, the captain is on the right, and the ensign or junior subaltern on the left of the front rank, the lieutenant in the rear, as also the drummer or pioneer in a third rank, at three paces' distance. In this formation companies are to assemble on their private parades, being sized from flanks to centre.

The

The company will be told off in sub-divisions, and four sections. If four officers are present, the captain takes the leading section, the next in seniority the rear section, the third in rank the third section from the head of the column, and the junior officer the second. The covering serjeant will cover the second file from the pivot of the leading section.—When there are but three officers, the covering serjeant will take the second section from the head of the column. The company will also be told by threes from the right, numbered 1, 2, 3.

Fig. A.

The four best-trained soldiers are to be placed in the front rank, on the right and left of each sub-division.

When thus formed, the company will be practised in

Opening, and Closing of } Ranks. (Sect. 27 and 28.)

Dressing { to the front, to the rear, in an oblique direction, } by the right and left;

and be exercised in the several motions of the firelock, as have been shewn in the preceding part.

Close order is the chief and primary order in which the battalion and its parts at all times assemble and form.—*Open order* is only regarded as an exception from it, and occasionally used in situations of parade and show. In close order, the rear rank is closed up to within one pace, the length of which is to be taken from the heels of one rank to the heels of the other rank, *a*. In open order, they are two paces distant from each other. (2)

Fig. A.

In order to distinguish the words of command given by the instructor of the drill (who represents the commander of the battalion) from those given by the commander of the company, or or its divisions, the commands of the former are in CAPITAL letters, those of the latter in *Italic.*

S. 43. *Marching to the Front.*

BY THE RIGHT (OR LEFT), MARCH

1. In the drill of the company, the person instructing must always consider it as a company in battalion, and regulate all its movements upon that principle; he will therefore, before he puts it in motion to front or rear, indicate which flank is to direct, by giving the word BY THE RIGHT, LEFT, OR CENTRE, MARCH, on which eyes will be directed full to the front, and the touch preserved to the named flank, or to the centre, as required.—Should the right be the directing flank, the commander of the company himself will fix on objects to march upon in a line truly perpendicular to the front of the company; and when the left flank is ordered to direct, he and his covering serjeant will shift by the rear to the left of the front rank, and take such objects to march upon. To MARCH on one object only, and to preserve a straight line, is an operation not to be depended on; the conductor of the company, before the word MARCH is given, will therefore endeavour to remark some distinct object on the ground, in his own front, and perpendicular to the directing flank: he will then observe some nearer and intermediate point in the same line, such as a stone, tuft of grass, *&c.*: he will move upon them with accuracy, and as he approaches the

the nearest of those points, he must, from time to time, choose fresh ones in the original direction, which he will by this means preserve, never having fewer than two such points to move upon. If no object in the true line can be ascertained, his own squareness of person must determine the direction of the march.

A person placed in the rear of a body can, more readily than if placed in its front, determine the line which is perpendicular to such front (vide S. 32.); and could we suppose ranks and files most perfectly correct, the prolongation of each file would be a perpendicular to the front of the body.

2. As the MARCH of every body, except in the case of inclining, is made on lines perpendicular to its front, each individual composing that body must remain perfectly square to the given line; otherwise he will naturally and insensibly move in a direction perpendicular to his own person, and thereby open out, or close in, according to the manner in which he is turned from the true point of his march. If the distortion of a single man operates in this manner, and all turnings of the head do so distort him, it may be easily imagined what that of several will occasion, each of whom is marching on a different front, and whose lines of direction are crossing each other.

The company, during its march in line, will occasionally be ordered to

Step out	vide Sect.	8.
Mark Time..........	———	10.
Step short	———	9.
Open and close ranks ..	———	33.
Oblique	———	36.
Diagonal March......	———	37.

S. 44

S. 44. *The Side Step.*

The *side* or *closing step* must also be frequently practised; it is very necessary and useful on many occasions, when halted, and when a very small distance is to be moved to either flank.—As, for instance, to open or close files; to join one division to, or open it from, another; to regain an interval in line; to move a whole battalion, or parade, 20 or 30 paces to a flank; to regulate distances between close columns before deploying; alterations made in this manner are imperceptible from the front, and better made than by facing and file marching.

TO THE RIGHT OR LEFT CLOSE. QUICK MARCH. HALT.	When the whole company is to close, at the word TO THE RIGHT CLOSE, the company officer takes one step to the front, and instantly faces about, the covering serjeant replacing him: On the word QUICK MARCH, the whole move together agreeably to the directions (in Sect. 11). On the word HALT, the company officer resumes his place, having stepped in the same manner as the men, but fronting them, and thereby assisted in preserving the direction.

At the word *Left close*, the officer will step nimbly out, and place himself in front of the left file of his company; and at the word *Halt*, he resumes his place in line by the rear.

S. 45. *The Back Step.*

STEP BACK, —MARCH,	The company must be accustomed from the halt, at the word STEP BACK—MARCH, to step back any ordered number of paces in the slow time and length, as it is an operation that may be frequently required from a battalion.

S. 46.

S. 46. *To form Three Deep.*

FORM THREE DEEP. MARCH.

Fig. B.

1. At the word MARCH, the rear rank man of the file, No. 2, takes one pace to the rear, with the left foot, and a pace to the right, with the right foot, so as to cover the rear rank man of the file, No. 1; at the same instant, file, No. 3, takes one pace to the rear with their left feet, and a pace to the right, so as to cover the front rank man of file, No. 2.

REAR FORM THREE DEEP. MARCH.

2. When three deep is formed facing to the rear, the word *March* is given, as above, and the instant the men cover, the whole face to the right about.

FRONT.

3. Two deep is re-formed from three deep in the following manner: Upon the word FRONT, the file, No. 3, steps up to the left of No. 2, the rear rank man of No. 2, steps nimbly up and covers his front rank man. If three deep has been formed to the rear, the whole come to *right about* on the word *Front*, and then instantly form up in the same manner.

The threes may close their intervals by the side step to the centre, or to either flank, and again open out, as may be required.

Marching to a Flank by Threes.

THREES RIGHT.

Fig. B.

4. The company receives the command THREES RIGHT, and the whole face to the right; the rear rank man of file, No. 2, forms on the right of file, No. 1, and file, No. 3, forms on the right of the front rank man of file, No. 2.

5. On

THREES LEFT. — 5. On the command THREES LEFT, the whole face to the left; the rear rank man of file, No. 2, forms on the left of file, No. 1, and file, No. 3, forms on the left of the front rank man of file, No. 2.

HALT, FRONT. — 6. Upon the word HALT, FRONT, the whole front and instantly form up two deep, by the rear rank man of No. 2, covering his front rank man, and file, No. 3, moving up to the left of No. 2.

The company must be frequently practised in these formations until the men are perfectly trained to form at the words THREES RIGHT, THREES LEFT, THREES INWARDS, THREES OUTWARDS, without pause or hesitation; and with a readiness equal to the execution of the commands applied to the different facings in file.

The threes must likewise be accustomed to turn on the march *Two deep* to the front and rear, preserving their perpendicular line, unless an order is given to retain their original formation.

Should the telling off by threes leave an odd file, it should form on the right of the left threes, so that in forming threes it retains its place at the usual distance between the two left sections of threes. When there are two men over the threes, they form a file in the same place; and when there are three, they form a single rank of threes: When four, they form a section of two file only; and when five, they form in the same manner, leaving the front rank man of No. 3 uncovered. The outer section of threes must always be complete.

S. 47. *To form Four Deep.*

F. C

FORM FOUR DEEP.

MARCH.

1. The company will be told off from the right by alternate files right and left; and when it is intended to form four deep, preserving the same front, the word FORM FOUR DEEP will be given, upon which the rear rank will step back one pace, and on the word MARCH, the left files will double behind the right files, by taking one pace to the rear with the left feet, and one pace to the right with the right feet, which forms four deep, leaving the intervals which the left files had quitted.

REAR.

FORM FOUR DEEP.

MARCH.

2. *When four deep is to be formed to the rear.* On the same caution the rear rank will step back as before, and on the word MARCH, the whole go to the right about, and the left files will then double as before, behind the right files.

RIGHT

FORM FOUR DEEP.

MARCH.

3. *When ground is to be taken to the right in the formation of four deep.* On the caution the rear rank steps back as before, and on the word MARCH, the whole face to the right, and the right files, at the same instant, form on the right of the left files.

LEFT

FORM FOUR DEEP.

MARCH.

4. *When ground is to be taken to the left.* The rear rank steps back as before, and on the word MARCH, the whole face to the left, and the left files, at the same instant, form on the left of the right files.

In

In all these formations the files preserve their proper order and place, and two deep is re-formed from each of them by the word *Front,* upon which the files move up to their respective intervals into line, the rear rank immediately closing on the front rank: the word *Halt* will precede the word *Front* if the company has been in movement.

In moving to a flank, file marching may be adopted, if necessary, by the files leading out in their proper order upon the word FORM TWO DEEP; and at the word RE-FORM FOUR DEEP, resuming their former places.

In the telling off for four deep, the odd files will be placed asexplained in the last section.

On the March in Line.

FORM FOUR DEEP.

5. *The company advancing or retiring in line, will form four deep on the march.* Upon the word FORM FOUR DEEP, the rear rank will mark time one pace, and the left files will double in rear of the right files, taking up the step from the front rank. In retiring, the front rank will mark time one pace, and the left files double as before, taking up the step from the rear rank.

Upon the word RE-FORM TWO DEEP, the left files will resume their places, and the rear rank will close to the front.

S. 48. *File Marching.*

LEFT FACE.

In marching by files, the commander of the company will lead the front rank; therefore, when the movement is by the left; on the word TO THE LEFT FACE, he, and his covering

QUICK MARCH. *Halt, Front.*

ing serjeant, will instantly shift to the left flank of the company by the rear; at the word QUICK MARCH, the whole step off together (vide Sect. 19); and on the word *Halt, Front,* the leader and his serjeant will return to their posts on the right in the same manner.

S. 49. *Wheeling from a Halt.*

RIGHT WHEEL, QUICK MARCH.

Halt, Dress.

In wheeling either forward or backward from a halt, the commander of the company, on the word RIGHT or LEFT WHEEL, moves out, and places himself one pace in front of the centre of his company: during the wheel, he turns towards his men, and inclines towards that flank which has been named as the directing or pivot one giving the word *Halt, Dress,* when his wheeling man has just completed the required degree of wheel: he then squares his company, but without moving what was the standing flank, and takes his post on the directing flank.

S. 50. *Wheeling forward by Sub-divisions from Line.*

BY SUB-DIVISIONS, RIGHT WHEEL.

1. On the caution by SUB-DIVISIONS, RIGHT WHEEL, the commander of the company places himself one pace in front of the centre of the right sub-division, at the same time the men on the right on the front rank of each sub-division face to the right.

At

QUICK MARCH. — At the word QUICK MARCH, each sub-division steps off in wheeling time, observing the directions given in Sect. 20 and 38. The commander of the company turning towards the men of the leading sub-division, and inclining to its left (the proper pivot flank),

Halt, Dress. — gives the word *Halt, Dress,* for both sub-divisions, as his wheeling man is taking the last step that finishes the wheel square; and instantly posts himself on the left, the pivot flank.—The serjeant coverer, during the wheel, goes round by the rear, and takes post on the pivot flank of the second sub-division. It is to be observed, that the commander of the company invariably takes post with the leading sub-division; therefore, when the company wheels by sub-divisions to the left, the commander of the company moves out to the centre of the left sub-division, and during the wheel inclines towards the right, now become the proper pivot flanks of the sub-divisions.

2. The company marching to the front may be wheeled into open column of sub-divisions on the moveable pivot, to either flank *without* halting; the instructor giving the word FORWARD, when the sub-divisions have wheeled square into column.

The *proper* pivot flank in column is that which, when wheeled up to, preserves the divisions of the line in the natural order, and to their proper front: the other may be called the *reverse* flank.

In column, divisions cover and dress to the proper pivot flank: to the left when the right is in front; and to the right when the left is in front.

S. 51. *Wheeling backwards by Sub-divisions from Line.*

CAUTION.

The company will also break into open column of sub-divisions by wheeling backwards. —When the right is intended to be in front: at the caution BY SUB-DIVISIONS ON THE LEFT, BACKWARD WHEEL, the commander of the company moves out briskly and places himself in front of the centre of the right sub-division.—The man on the left of the front rank of each sub-division at the same time faces to the right.

QUICK MARCH

On the word QUICK MARCH, each sub-division wheels backward, as directed in Section 22, and Section 39. During the wheel, the commander of the company turns towards his men, inclining at the same time to the left, or pivot flank, and on completing the wheel gives the word *Halt, Dress,* to both divisions: he and his covering serjeant then place themselves on the left flanks of their sub-divisions.

Halt, Dress.

S. 52. *Marching on an Alignement, in Open Column of Sub-divisions.*

The company having wheeled backwards, by sub-divisions from line, (as directed in the foregoing Section), and a distant marked object in the prolongation of the two pivot flanks being taken; the commander of the company, who is now on the pivot flank of the

the leading sub-division, immediately fixes on his intermediate points to march on, (Vide S. 43.) On the word MARCH, given by the instructor of the drill, both divisions step off at the same instant; the leader of the first division marching with the utmost steadiness and equality of pace on the points he has taken; and the commander of the second division preserving the leader of the first in the exact line with the distant object; at the same time he keeps the distance necessary for forming from the preceding division, which distance is to be taken from the front rank. —These objects are in themselves sufficient to occupy the whole attention of the leaders of the two divisions; therefore they must not look to, nor endeavour to correct the march of their divisions, which care must be entirely left to the non-commissioned officers of the supernumerary rank.

MARCH.

S. 53. *Wheeling into Line from Open Column of Sub-divisions.*

1. The company being in open column or sub-divisions, marching on the alignement, receives the word HALT from the instructor of the drill; both divisions instantly halt and the instructor sees that the leaders of the divisions are correct on the line in which they have moved; he then gives the word (supposing the right of the company to be in front)

HALT.

LEFT WHEEL INTO LINE.

front) by sub-divisions LEFT WHEEL INTO LINE; on which the commander of the company goes to the centre of his sub-division, the two pivot men face to their left exactly square with the alignement, and a serjeant runs out and places himself in a line with them, so as to mark the precise point at which the right flank of the leading sub-division is to halt, when it shall have completed its wheel.—At the word QUICK MARCH, the whole wheel up in wheeling time; during the wheel, the commander of the company, turning towards his men, inclines to the wheeling flank, and gives the word *Halt, Dress,* at the moment the wheel of the division is completed: the commander of the company, if necessary, corrects the internal dressing of the company on the serjeant and pivot men: this dressing must be quickly made, and, when done, the commander of the company gives the word *Eyes front,* in a moderate tone of voice, and takes post in line, as directed in Sect. 42.

QUICK MARCH.

Halt, Dress.

Eyes front.

2. The company may be wheeled into line on the march on the moveable pivot, receiving from the instructor either the word FORWARD, or HALT, DRESS, when the wheel is complete.

In all wheels of the divisions of a column that are to be made on a halted pivot in order to form line, the flank firelock of the front rank on the hand wheeled to is such pivot, not the officer who may be on that flank, and whose business is to conform to it.

All wheelings by sub-divisions, or sections, from line into column,

column, or from column into line, are performed on the word given by the commander of a battalion, when the whole of a battalion is at the same instant so to wheel; or on the word given by the commander of the company, when companies singly, or successively, so wheel; they are not to be repeated by the leaders of its divisions.

When the company is in open column of subdivisions, it must be occasionally practised to wheel into line upon the reverse flank of the column; and in shewing a front line also to that reverse flank by wheeling backwards on the regular pivots.

S. 54. *In Open Column of Sub-divisions wheeling into an Alignement.*

Halt, Wheel.
Halt, Dress.
March.
Halt, Wheel.
Halt, Dress.
March.

When the company marching in open column of sub-divisions arrives at the ground, where the wheel is to commence, the leading division receives the word *Halt, right,* or *left,* wheel, from its commander; on which the rear rank, if at one pace distance, locks up; the flank front rank man alone halts, and faces into the new direction, while the others quicken their pace to the wheeling time, and regulate their step by the outward man (to whom they have turned their eyes), until the wheel is completed.—He then gives the word *Halt, Dress,* and immediately the word *March,* and moves on so that its rear rank may not occasion even a momentary stop to the division behind it, which at that instant receives the word *Wheel,* then *Halt, Dress,* and finally *March.* Any pause is to be carefully avoided

The officer conducting the leading (and every other) division of the column in march, on any given point or object where it is to wheel into a new direction, and to its proper halted pivot; always stops at that point, or object, close on his own outward hand, and gives the word WHEEL, when the front rank of his division has taken ONE pace beyond such object; he thus allows space for his own person (when the wheel is finished) to move on the new direction of march.

But if the proper pivot flank is to be the wheeling one, each commander of a division gives his word *Wheel*, as he successively arrives at such a distance from the point on which he has moved, as that at the completion of the wheel, his division may *halt* perpendicular to the new line, but with the given point, of course, behind the proper pivot; and that he also in his own person be on the new direction, prepared to give his word *March*, and to proceed.

The sub-divisions must take care that they continue their march correctly upon the point where the leading one wheeled, and that they do not shift to either flank, which, without much attention, they are apt to do.

In this manner the sub-divisions succeed each other: and if the words of command be justly given; no stop made on arriving at the wheeling point; and the wheels performed at an increased time and step; no extension of the column will take place, but the just distances between the divisions will be preserved at the word *Halt*.

S. 55. *In Open Column of Sub-divisions entering into a new Direction on a moveable Pivot.*

Right Shoulders forward.

Forward.

The commander of the leading sub-division, when it arrives at the new direction, will give the word *Right* (or left) *Shoulders forward* (Vide Sect. 23), and when his sub-division has wheeled square to that direction, he will give the word *Forward.*—The leader of the second sub-division, when he arrives at the ground where the first began to change its direction, will give the same words, following the exact track, and always preserving his distance from the division in his front.

S. 56. *Countermarching.*

The company, when it is to countermarch, must always be considered as a division of a battalion in column; the instructor of the drill will therefore, previous to his giving the caution to countermarch, signify whether the right or left is supposed to be in front, that the commander of the company, and his covering serjeant, may be placed on the pivot flank before such caution is given, as it is an invariable rule in the countermarch of the divisions of a column by files, that the facings be made from the flank, then the pivot one, to the one which is to become such.

Countermarch by Files.

RIGHT (OR LEFT) FACE.

1. On the word, RIGHT, or left, FACE, the company faces, the commander of it immediately goes to the other flank, and his covering serjeant faces to the right about.—At the word

QUICK

QUICK MARCH.

QUICK MARCH, the whole, except the serjeant coverer, step off together, the company officer wheeling short round (to his left, if he has shifted to the right of the company; or to his right, if he has shifted to the left of it); and proceeds, followed by the company in file, till he has conducted his pivot front rank man close to his serjeant, who has remained immoveable; he then gives the words *Halt, Front,* and *Dress,* squares, and closes his company on his serjeant, and then replaces him.

Halt, Front, Dress.

Countermarch by Ranks.

RIGHT AND LEFT FACE.

2. On the word FACE, whether the right o left is in front, the front rank faces from the pivot, the rear rank to it; officers place themselves on the outward flank of their serjeants facing inwards, and the covering serjeants go to the right about.

RIGHT OR LEFT. COUNTER-MARCH. QUICK MARCH *Halt, Front, Dress.*

The whole step off together, the two ranks severally wheeling in single file, till the leading file of the front rank comes close to the covering serjeant; they then receive the word *Halt, Front, Dress,* from the officer who replaces the serjeant.

Countermarch by Files from both Flanks.

3. The company standing in distinct sub-divisions (an officer and a serjeant coverer on the outer flank of each sub-division, with a central serjeant marking the point of separation between the two) receives the caution to countermarch

countermarch and change ground from both flanks, on which the covering serjeants on the flank of each sub-division exchange places, and face to the future front, in the line of the rear rank, and the central serjeant goes to the right about.

SUB-DIVISIONS OUTWARD FACE. — The sub-divisions face with their respective officers.

RIGHT COUNTER-MARCH. QUICK MARCH. *Halt, Front, Dress.* — The whole step off together, each sub-division wheeling short round to the right, and proceed as directed for the countermarch by files; each sub-division receiving the word *Halt, Front, Dress,* from its own officer, as soon as they have respectively changed places.

All countermarches by files necessarily tend to an extension of the files; unity of step is therefore absolutely indispensable, and the greatest care must be taken that the wheel of each file be made close, quick, and at an increased length of step of the wheeling man, so as not to retard or lengthen out the march of the whole.

Companies, or their divisions, when brought up in file to a new line, are not to stand in that position, till the men cover each other minutely; but the instant the leading man is at his point, they will receive the word *Halt, Front,* and in that situation close in and dress correctly.

The countermarch from both flanks of the company is principally applicable to the countermarch of a double column of sub-divisions, in which the wings change places: On other occasions the countermarch by subdivisions is performed by their wheeling, the reverse sub-division going to the right about, and coming to the front when halted.

S. 57.

S. 57. *Wheeling on the Centre of the Company.*

The company must be accustomed to wheel upon its centre, half backward, half forward, and to be pliable into every shape, which circumstances can require of it; but always in order, and by a decided command.

The words of Command are,

COMPANY, ON THE CENTRE { RIGHT, / LEFT, / RIGHT ABOUT, / LEFT ABOUT, } WHEEL.

When the wheel is to the right, or right about, the right half company wheels backward, and the left forward.—In this case the right hand man of the left sub-division is the pivot man; he faces to his right, or right about, and the covering serjeant springs out and alignes himself with him, but to the flank which is to become the pivot.—The reverse will take place, when the wheel is to be made to the left, or to the left about. —The left hand man of the right subdivision is then the pivot man, who will face to his left, or left about; the covering serjeant alignes himself with him, as in the wheel to the right.—On the word MARCH, the whole move off together in the wheeling time, regulating by the two flank men, who, during the wheel, preserve themselves in a line with the centre of the company; as soon as the required degree of wheel is performed, the commander of the company gives the word *Halt, Dress,* and instantly squares it from that flank, on which he himself is to take post.

MARCH.

Halt, Dress.

S. 58. *Diagonal March.*

The instructor of the drill will have the diagonal march frequently practised, in company, and in sub-divisions; (vide S. 36, 37.) He will see that the rear rank locks well up, and covers exactly; that the exact distances are preserved between the files: and that the pivots, or outward files, march in the direct line to which they have faced, the others conforming to them.

S. 59. *Increasing and Diminishing the Front of an Open Column halted.*

Increasing.

Pl. II. Fig. 1—A

FORM COMPANY.

RIGHT SUB-DIVISIONS, RIGHT ABOUT THREE QUARTERS FACE; QUICK MARCH

Halt, Front, Dress.

The company standing in open column of sub-divisions (suppose the right in front) receives from the instructor of the drill a caution to FORM COMPANY; upon which the covering serjeant will run out to mark the reverse flank, (*r*) the instructor will instantly order, RIGHT SUB-DIVISIONS, RIGHT ABOUT THREE QUARTERS FACE; QUICK MARCH; and the reverse file will march straight to the covering serjeant.—When the sub-division has obliqued so as to gain the line of the left sub-division, the commander gives the word *Halt, Front, Dress;* and takes post on the left, the pivot flank of the company. The men front to the left from the right about three quarters face, vide Part I.

On

Diminishing.

Fig. D.
Pl. II. Fig. 1—B.
FORM SUB-DIVISIONS.

On the cautionary command from the instructor of the drill to FORM SUB-DIVISIONS, the commander advances to mark the point where the left flank of the right sub-division is to rest (*c*). The instructor of the drill, while the commander is advancing to that point, orders, RIGHT SUB-DIVISION, LEFT HALF FACE, QUICK MARCH; and the file of the inner flank of the right sub-division marches straight to the commander, and when it shall reach him, the sub-division receives the word, *Halt, Front, Dress.*—The commander remains on the left flank of the right sub-division (*c*), and his serjeant on that of the left (*s*).

RIGHT SUB-DIVISION, LEFT HALF FACE. QUICK MARCH. *Halt, Front. Dress.*

It is to be observed, both in increasing and diminishing the front of an open column halted, that upon the usual caution, the reverse file of the pivot sub-division falls back one pace (*b*) to leave room for the flank of the reverse sub-division, and upon the word *Halt, Front,* it resumes its place.

The sub-divisions or sections on the reverse flank, must always double in front of the pivot sub-divisions, or sections.

Fig. D. Thus, when the right is in front, the doubling will be in front of the left division: and, when the left is in front, it will be in front of the right division. When the front of a column is increased, the front sub-divisions, or sections, make a three quarters face to the right, and move by the diagonal march to the reverse flank; so that when the right is in front, the diagonal movement will be to the left and rear, and the reverse when the left is in front.

S. 60. *Increasing and diminishing the Front of an Open Column on the March.*

Increasing.

Fig. 2—A.

FORM COMPANY. RIGHT SUB-DIVISION, RIGHT TURN.

The company marching in open column of sub-divisions (*a a*) suppose right in front), receives from the instructor of the drill the cautionary command, FORM COMPANY, RIGHT SUB-DIVISION, RIGHT TURN; the men at the word *Turn*, lengthen their pace to 33 inches, and when the division has cleared the extent of its own front (*c*), the left sub-division (*e*) which has continued to march with the utmost steadiness, will have gained its inner flank; and then the commander of the company gives the word *Front turn* to the right sub-division which moves on in line with the left sub-division, and takes post on the pivot flank of the company towards which he has been moving.

Front turn.

Diminishing.

Fig. 2—B.

FORM SUB-DIVISIONS.

When the instructor of the drill gives the caution to FORM SUB-DIVISIONS, the commander of the company, advances to the proper distance in front (*o*), the instructor then gives the word, RIGHT SUB-DIVISION, LEFT HALF TURN, and it instantly moves off at the *Double March*, if the column has been moving in quick time; and when the inner file of the reverse subdivision shall reach the commander, he gives the words *Front turn, Quick,* and the sub-division takes up the step at which th pivot divisions of the column are moving. If the

RIGHT SUB-DIVISION, LEFT HALF TURN.

Front turn, Quick.

Slow. — the column has been marching in slow time, the reverse divisions double up in quick time, and resume the step of the pivot division at the word *Slow*.

Upon the usual caution in diminishing as before, the reverse file of the pivot sub-division will mark time one pace, to leave room for the diagonal advance of the right sub-division; and when the flanks are clear, it will resume its place.

The instructor of the drill must take particular care that the pivot divisions continue their march at the regular time and length of pace, and that the exact distances between the divisions are accurately preserved.

S. 61. *The Company in Open Column of Sub-divisions to pass a short Defilé, by breaking off Files.*

BREAK OFF 3 FILES.

Three files, left turn, right wheel.

The company is supposed in open column of sub-divisions, with the right in front; when the leading division is arrived within a few paces of the defilé, it receives from the instructor of the drill an order to break off a certain number of files (suppose three).—The commander of the leading division instantly gives the words, *Three files on the right, left turn, right wheel*; the named files immediately turn to the left, and wheeling to the right follow in file in rear of the right flank of the sub-division.—When the second sub-division comes to the spot where the first division contracted its front, it will receive the same words of command from its own leader, and will proceed in like manner.

Should

Two files, left turn.

Should it be required to diminish the front of the column one or two files more, the commander of the leading division will, as before, order the desired number of files to *turn:* on which those already in the rear will incline to the left, so as to cover the files now ordered to break off, and which turn to the left and wheel to the right in the manner already prescribed.

In this movement, the files in the rear of the sub-divisions must lock well up, so as not to impede the march of the succeeding division.

Three files to the front.

As the defilé widens (or the instructor of the drill shall direct) the commander of the leading sub-division will order files to move up to the front, by giving the word, *One, two, or three files to the front:* on which the named files turn to their front (the right), and lengthening their pace, march up, file by file, to the front of their sub-division, and immediately resume the march.—Those files which are to continue in the rear will oblique to the right lengthening also their step, till they cover, and are closed up to the right flank of their sub-division.

It is to be observed that in passing a defilé, the files always break off from the reverse flank.

S. 62. *Forming Company, Sub-divisions, or Sections, from the Flank March of Threes, Right in Front.*

FRONT FORM COMPANY (OR SUB-DIVISIONS.)

Upon the word FRONT FORM COMPANY, the leading man will mark time, the remainder will turn their bodies a whole face to the left, and wheel to the right, and that instant will form two deep: If the company are to keep on the march the word FORWARD is given.

The same rule to be observed, if left in front, with the difference of turning their bodies to the right and wheeling to the left.

THREES RIGHT LEFT WHEEL,

THREES LEFT RIGHT WHEEL.

If the flank march is again to be resumed, the word is given THREES RIGHT, LEFT WHEEL, if the column is right in front; and if the column is left in front, the word is given THREES LEFT, RIGHT WHEEL.

S. 63. *Forming Company, Sub-divisions, or Sections, from File Marching.*

FRONT FORM COMPANY, SUB-DIVISIONS, OR SECTIONS.

At this word of command, the leading file marks time, the remainder turn their bodies a whole face to the left, and wheel to the right, looking to the outward flank and feeling inwards (that is to say) if right in front, turn to the left, and if left in front, turn to the right.

As soon as the quarter circle is complete, the word FORWARD is given, if the march is to be continued.

At

ON THE LEADING FILE TO THE RIGHT FORM COMPANY.

At this word of coammnd, the leading file will halt and face to the right, the remainder of the company form on the lef of the right file, by files in succession.

ON THE LEADING FILE TO THE LEFT FORM COMPANY.

The same rule is to be observed in forming to the left, with this difference, the leading file will halt and face to the left, and the remainder will form on the right by files in succession.

ON THE LEADING FILE TO THE RIGHT ABOUT, FORM COMPANY.

At this word the leading file halts and faces to the right about, the remainder of the company march on in file and form on the left of the leading file in succession, and halt as they come into the line.

ON THE LEADING FILE TO YOUR LEFT ABOUT, FORM COMPANY.

The same rule is to be observed in this, with the difference of the leading file facing to the left about, and the remainder forming on the right.

S. 64. *To form to either Flank, from Open Column of Sub-divisions.*

HALT, LEFT WHEEL INTO LINE, MARCH.

The company marching in open column of sub-divisions, to form to its left, receives the words, HALT, LEFT WHEEL INTO LINE, —MARCH, &c., and proceeds as has already been shewn in S. 53.

To

RIGHT FORWARD FORM COMPANY.	To form the company to its right flank, the instructor of the drill gives the cautionary word of command, TO THE RIGHT FORM THE COMPANY; on which the commanders of the several divisions shift to the right flank, and the commander of the leading sub-division instantly gives the word to his
Left Shoulders forward. Halt, dress.	division, *Left Shoulders forward.* When it has wheeled square, he orders, *Halt, dress;* and dresses it on the intended line of formation.—The commander of the other sub-division, on the leading one being ordered to
Left half turn.	wheel, gives the word, *To the left half turn,* and gradually inclines, so as to be able to march clear of the rear rank of the sub-division forming: this being well effected, the
Front turn.	word *Front turn,* will be given to the sub-division, and it will move on in the rear of the one formed.—When the second sub-division is arrived at the left flank of the first,
Left Shoulders forward. Halt, Dress.	its commander gives the word, *Left Shoulders forward,* then *Halt, dress;* on which the division moves up into the line with the one formed: and its commander from the left of his first division, dresses his own on the given flank point, as quickly and as accurately as possible, and resumes his proper company place.

This formation to a flank may also be performed by a wheel of sub-divisions on their halted pivots.

S. 65. *The Company moving to the Front, to gain Ground to a Flank, by a March in Echellon, by Sections.*

In the drill of the company, when the soldier is completely formed, he may be taught to march in echellon by sections. This is a very useful movement for a battalion, or any large body moving in line, or column, that is required to gain ground to a flank; and it may be substituted instead of the oblique march.—It will be performed in the following manner:

SECTIONS, RIGHT. FORWARD.	The company, marching to the front, receives the word by SECTIONS RIGHT; the right hand men of the front rank of each section turning in a small degree to their right, mark the time two or three paces, during which the sections are wheeling on their pivot men; and on the word *Forward*, the whole move on direct to the front that each section has now acquired: and the company continues its march in echellon.
FORM COMPANY. FORWARD.	On the word FORM COMPANY, the pivot men mark the time as before, turning back in a small degree to the left, the original front, and the sections instantly wheel backward into line; on the word FORWARD, the whole advance in line.

S. 66. *Marching.*

The company must be well drilled in the different degrees of marching, viz., Slow—Quick—and the Double March, until the men have acquired the utmost precision in these movements.—It ought to be much practised in the double march, in all the formations of *threes*, as well as in the company formed in line—in column of sub-divisions, sections, &c. &c., that march being essential in the interior formations of the battalion.

S. 67. To form the Rallying Square.

Fig. E.

FORM THE RALLYING SQUARE.

The instructor of the drill having caused the recruits to fall out and disperse to a certain distance, will give the word FORM THE RALLYING SQUARE at the same time placing himself facing the supposed enemy; the recruits hasten to the person so posted, fixing bayonets and shouldering their arms as they reach him. The two first who join him form on his right and left, facing outwards. (1). The three next place themselves in front of those posted, (2) and three others to the rear facing to the rear, thus forming a square of three. (3) The instructor will cause the next four men to take post at the several angles; (4) and others as they come up will complete the different faces between these angles, which will form a square of five. If the front rank of either of the faces of the square should be incomplete, the instructor will fill the vacancies from the rear rank.

2. A square thus composed of twenty-four men (besides the person who is to rally) and formed two deep, may be augmented to a square of *seven*, three deep, by four more men

takgin

Fig. E.

taking post at the angles, and others coming up to complete the faces as before; the square will then consist of forty-eight men, and may be augmented in the same manner to a square of nine, four deep, by the angles being occupied by four more men, and the faces filled up as before; and the square will then be composed of eighty men.

3. *If a mounted officer is to rally the dispersed men*—He will give the words *Form the Rallying Square*, and five men will form in his front, five in the rear, and three on the flanks; the rest of the formation proceeds as before.

THE SQUARE WILL MOVE TO THE FRONT, REAR, RIGHT, OR LEFT.

INWARDS FACE.

QUICK MARCH.

4. *When the square is to march.*—In order to move with the necessary regularity previously to putting the square in motion, the instructor will cause the faces to be dressed; and after the caution that THE SQUARE WILL MOVE TO THE FRONT, REAR, RIGHT, OR LEFT, he will give the words INWARDS FACE, and it will face to the named face of the square, and step off accordingly at the word QUICK MARCH.

HALT.

PREPARE TO RESIST CAVALRY.

READY.

5. *To resist cavalry.*—Upon the word HALT, the square will halt and face outwards, and when it is to PREPARE TO RESIST CAVALRY;—upon the word READY, the front rank only (if the square is two or three deep) will kneel; if four deep, the two front ranks will kneel, and plant their bayonets. If ordered

FIRE. dered to fire, the standing ranks only will commence an independent fire, bringing the firelock gradually up to the present.

In this manner small dispersed parties, from eight to eighty men, may be formed to resist an attack of cavalry in an open country, where either from fatigue, or other causes, soldiers may have separated from the column of march. The formation must be frequently practised in squad and company drill, so that soldiers, thoroughly instructed, may always be enabled to protect themselves upon an emergency.

In pursuance of the foregoing instructions, and on the principles they contain, every company of a battalion must be frequently exercised by its own officers, each superintending a rank, or an allotted part of the whole movement. On a space of 70 or 80 yards square, every movement and formation may be practised that is necessary to qualify it for the operations of the battalion.

Officers must be instructed in the exercise of the sword; and they must be habituated to give their words of command with energy and precision.—Every officer, on first joining a regiment, must be drilled and exercised until he shall have a perfect knowledge of all the detail of drill required from a soldier. He cannot be considered capable of instructing the men under his command, nor be permitted to take the command of a company in the battalion, until he is master of all these points.

Squads of officers must be formed, and exercised by a field officer: They must be marched in all directions; and in proportion to the number of files in a division, they must learn, accurately to judge, the distances necessary for each, and to extend that knowledge to the front of greater bodies. They must

must acquire the habit of readily ascertaining, by the eye, perpendiculars of march, and the squareness of the wheel.

An officer ought to know the post which he should occupy in all changes of situation,—the commands which he should give,—and the general intention of a required movement. He should be master of the principles on which each formation is founded; and aware of the faults that may be committed, in order to avoid them himself, and to instruct others. —These principles are in themselves so simple, that moderate reflection, habit, and attention, will soon shew them to the eye, and fix them in the mind; and individuals, from time to time, when qualified, must be ordered to exercise the battalion, or its parts.

The complete instruction of an officer enlarges with his situation, and at last takes in the whole circle of military science: —From the variety of knowledge required of him, his exertion must be unremitting, to qualify himself for the progressive situations at which he may arrive.

END OF PART SECOND.

PART III.

GENERAL PRINCIPLES

FOR THE MOVEMENTS OF A BATTALION.

(Plate II.)

S. 68. *Commands.*

1. All words of command, must be given short, quick, and loud, so as to be heard and understood from right to left of a battalion in line: or from front to rear of a battalion in column. And, indeed, every officer must be accustomed to give his words of command, even to the smallest bodies, in the full extent of his voice, and in a sharp tone, in order to be heard by the leaders of other corps who are dependent on his motions.—The confidence and prompt obedience of the soldier, can only be in proportion to the firm, decided, and proper manner in which every officer gives his orders.

2. In the midst of surrounding noises, the eye and the ear of the soldier should be attentive only to his own immediate officer; the loudness of whose commands, instead of creating confusion and unsteadiness, ought to give confidence in the hurry of action.

3. The field officers and adjutant of the battalion are at all times mounted, in order to give ground in movements, to cor-

rest

rect mistakes, to circulate orders, to dress pivots, when they ought to cover in column in a straight line, and to take care, when the column halts, that they are speedily adjusted before wheeling up into line. No dismounted officer can possibly perform the duties required of those who fill such situations.

S. 69. *Degrees of March.*

General Intention. 1. All military movements are intended to be made with quickness consistent with order, regularity, and without hurry or fatigue to the troops.—The uniformity of position, cadence, and length of step, produce equality and freedom of march, on which every thing depends, and to which the soldier must be carefully trained.

2. The different degrees of march have been already detailed in the first and second parts; and the soldier must be trained and constantly practised in them without drum or music, in order that by habit alone he may be taught to acquire the given length of step. Troops must alone trust to this equal and unvaried cadence for the preservation of their line under all the various and local circumstances of service which call for its accuracy.

Slow March. 3. The slow step, hitherto called the *Ordinary March*, need only be applied to the purposes of parade, upon some occasions to the march in line, or when required for any special object.

Quick March. 4. The *Quick March* is the ordinary pace to be applied to all general movements of battalions, or greater

greater bodies, in column or line: The *Double March* to the movements of the divisions of a battalion; that is, to the interior movements of the divisions of a battalion when forming on any of its fixed or particular parts. The *Double March* cannot be applied to the movements of greater bodies, without exhausting the men, except upon peculiar occasions for a short distance; as in a charge, or where a post, or position is to be seized, or the wheel of a column to be performed. Commanding officers of battalions will use a discretion in the application of the *Double March*, for interior movements, (taking care that it is never adopted unnecessarily when the men are in heavy marching order); and it may sometimes be expedient that the divisions should be put in motion in quick time, and allowed to disengage before the word *Double March* is given. In echellon movements the *Double March* may be safely applied to sub-divisions or sections, when used in rapid formations from line to square, and from square to line, or for quickly throwing back or forward a flank; but not to the march of companies or strong divisions in echellon changes of front or position. (Vide S. 78, No. 17 and 18.) Much of this, however, will depend upon the degree of perfection, at which the drill of a battalion has arrived, and upon the attention and intelligence of division officers. But it must be kept in the constant recollection of commanding officers, that perfection in the *Double March* can never be attained (and therefore it must never be practised) until the men are accurately placed in the position of soldiers, and thoroughly instructed in all movements at *Quick Time*.

Double March.

5. A careful attention to the position of the soldier

as

as directed in S. 1, and the additional impetus given to the body by the accelerated motion of the *Double March*, will enable him to take a lengthened step without fatigue to himself: The march is thus rendered effective; and great pains must be taken to habituate the soldier to the full pace of 36 inches (Vide S. 17.) If this be not carefully attended to, the men will get into the habit of a short trot, which would defeat the obvious advantages of this degree of march.

The diagonal march. Fig. D.

6. The diagonal march, as explained in S. 37, may be adopted with great advantage upon most of the occasions to which the oblique march has hitherto been applied. In all oblique movements of small divisions, it has the same effect proposed by the oblique march, without being liable to the inaccuracies which must be, more or less, inseparable from this step at all times.

Applicable to increasing or diminishing the front of columns. Fig. 1—B. Pl. I. Fig. D.

7. The diagonal march may be applied to the increase or to the diminution of the front of columns, at the halt, or in movement:—When the front is diminished, the sub-divisions that are to double up, make a half face towards the pivot flanks; and the files on their inner flanks (*a a*) take diagonal lines of direction, on making this half face, which cut the line of company pivots, (*o o*) where the pivot flanks of the reverse sub-divisions are to rest at their proper distances (*c c*).——Thus the files of the reverse sub-divisions are respectively placed at an angle of 45° with the perpendicular direction (*d d*) of their original front; and if the two flank files of each march straight forward along the opposite sides (*e p* and *c. a.*) of the parallelograms, and if the other files preserve their respective echellon or diagonal lines of direction, the subdivisions

subdivisions (*a p*) will be brought without difficulty to their proper position (*c e*) in front of the pivot subdivisions, either at the halt or on the march.—(Vide S. 59.) Columns of march, or manœuvre, may also take ground to a flank by the diagonal march.

Oblique march. 8. The instruction for the oblique step is still necessary, however, in the drill of the recruit, for the obvious purpose of giving him the essential habit of moving obliquely without affecting the square position of his body to the front; and this is highly useful to prevent crowding, or opening out, and to preserve correctness in line marching.

S. 70. *Marching in Line.*

1. The march in line, either to front or rear, is the most important and most difficult of all movements, and requires every exertion of commanding officers, and every attention of officers and men for its true attainment. The essential points to be observed, are, the perpendicular direction of the march to the front of the battalion as then standing;—the perfect squareness of shoulders and body of each individual;—the light touch of the files;—and the accurate equality of cadence and step, given by advanced serjeants.

Directing serjeanis. 2. Every individual should therefore be well prepared for this movement. But to ensure its correct execution, three or more directing (or colour) serjeants must be trained and formed in the centre of the battalion between the colours; upon whose exact cadence

cadence, step, squareness of body, and precision of movement, dependence can be placed. A serjeant is also to cover them in the supernumerary rank.

Fig. 3. 3. At all times when the battalion is formed in line, and directed to advance, the central serjeant of the three who lead the line, takes six paces to his front and halts. His serjeant coverer, who has taken his place in the line, covers him correctly, and the latter is again covered by the serjeant-major, six or eight paces in rear of the line (*a*).

4. These three points being corrected by the adjutant, or a mounted officer, a fourth point in prolongation is easily obtained. If a distinct and visible object should present itself in the true line, the mounted officer will order the directing serjeant to march upon it (*t t*). If this be not the case, the directing serjeant, (after being assured by the mounted officer that he himself is perfectly and squarely placed,) will, by casting his eyes down the centre of his body, from the junction of his two heels, take up and prolong a line perpendicular to himself and to the battalion; for this purpose he will observe and take up any accidental small point on the ground within 100 or 150 paces. Intermediate ones cannot be wanting, and their renewal is easy, as he successively approaches them in his march.

5. These preparatory arrangements being made, the mounted officer will give the word "*Steady*," and the other two centre serjeants will immediately move out, and align themselves upon the one already posted.

6. When

Fig. 3. 6. When the battalion advances, (A) the serjeant-major will remain steady in the rear of the line, for 20 or 30 paces (*a*), to ascertain under the direction of the mounted officer, the squareness and correctness of the line of march. If no waving or crowding in the line appear, the direction is certainly true, and the serjeant-major will then follow in the rear, covering as before.

7. The mounted officer will occasionally superintend the whole; but so long as these three points cover directly on the distant visible object, or on the intermediate ones, taken up successively by the directing serjeant in his own person, the correct squareness of the battalion may be depended upon.

To pass obstacles when marching in line. Fig. 3—B.

8. The march in line is generally adopted where the country is open, but in certain cases, where partial obstacles may offer, they can be avoided, by forming threes to the right or left and then right or left wheel to follow the formed flank of the line. The formation in this manner will increase as the obstacle increases; but as it diminishes the threes will successively form up, until the whole are again in line. Where the obstacles are of small extent, but frequently occurring, this mode is the most ready that can be applied in either advancing or retiring. Where they are of greater extent an entire column formation should be adopted.

Retiring. 9. In marching to the *rear*, the battalion must cover its proper extent of ground. The rear rank men must avoid closing their files more than usual, otherwise the front men, who are in general larger, will be be crowded in their ranks: Music, drums, supernumerary officers, &c., will take care to march with exactness, not to interrupt, but rather to assist the battalion. The battalion is not to *face* about, till every thing is prepared for its instant *march*, and its *halt, front,* is one command; when retiring, therefore, it never unnecessarily stands faced to the rear.

Flank companies.

10. When the flank companies of a battalion are wheeled backward, and faced outward, in order to cover its flank, such companies, if during the retreat they march by threes or in file, will take particular care to move in the same direction as the battalion,
Fig. 4. and not impede its progress. When the battalion fronts, those companies will face outward, and always recollecting that their immediate business is to cover the flanks, they will regulate their position and movements by those of the battalion. When marching A, they move perpendicular to the line of the battalion. When fronted B, they make an angle with it of about 45°, according to the apparent circumstances that threaten.

S. 71. *Wheeling.*

1. The manner of performing the wheels of a division is sufficiently explained in the first and second Parts, (S. 29 and 49). It is shewn that they are made on halted and moveable pivots; and it may only be necessary to add an explanation of the circumstances under which each description of wheel is respectively adopted.

2. Wheels

On a halted pivot. 2. Wheels are made on a halted pivot from line into column, and from column into line; and they may also be made in open column of march or manœuvre, when the change of direction is at a right angle (*a b d*) Fig. 5. with the line (*a b*) on which a column has been moving; and when of course the wheel of the quarter circle is necessary. When the head of a column thus changes its direction on a halted pivot, and the divisions successively wheel the quarter circle (*c*), the column must necessarily be formed at full distance. If it is formed at only half, or quarter distance, the succeeding divisions must arrive at the wheeling point before those in their respective fronts can have completed the wheel,—and thus crowding and confusion must ensue;—for, although the arc of the quarter circle (*c*), which a division describes in its wheel, is nearly one half more than the extent of its front,—yet, by the use of the wheeling pace, and lengthened step on the outer flank, the wheel is completed, at open distance, in sufficient time to prevent the crowding of divisions upon each other. But as the arc of the quarter circle so much exceeds the space between the divisions of column at half, or quarter distance, there is no pace, however rapid, that could accomplish the wheel before the succeeding division would reach the wheeling point; and consequently the divisions would crowd upon each other, and confuse and derange the march.

On a moveable pivot 3. The principle of the moveable pivot, therefore, (Vide S. 23.) must always be applied to the wheel of divisions in column at half, and quarter distance: And in column of route and manœuvre, the new direc- Fig. 6. tion should be taken at an angle (*abc*), so obtuse with the

the former line of march (*ab*) as not to require more han a wheel of one-eighth (*d*) (for half distance), and
Fig. 7. one sixteenth (*z*) (for quarter distance,) of the circle; otherwise, upon the principle above explained, the arc of the circle to be wheeled would so much exceed the distance between divisions, that crowding must be the consequence. If the change of direction in such columns is to be made on the moveable pivot, at an angle more acute than the above proportions, or at a right angle with the line of march, it must be performed gradually, (unless performed by battalions at once, vide S. 77. No. 5.)—the word "*Forward*" being always given when the wheel of the above-named extent has been completed; and then the wheel to be repeated, if necessary, when a few paces more advanced. And if the leaders of each division follow with scrupulous attention, the changes of direction of the division in their front—giving the word "*Right (or left) Shoulders forward*" at the same spot, and then "*Forward*" at the moment when the division is square with the one preceding, the distances, and the whole regularity of the march, will be preserved (*z z z z*). It may be observed, however, that a column of very small divisions may follow and conform to the windings and turnings of the head, without repeating the word, *Shoulders forward*, which should be given by the head division upon all occasions.

4. Successive wheels of the quarter circle into a new alignement, may be made on the moveable as well as the halted pivot, when the divisions are at full distance; and upon the principle laid down in S. 23, they will be found equally accurate with the halted pivot, in all the essential points of covering, dis-

tances

stances, &c., provided the leading division shall receive the word *Forward*, the very moment the quarter circle is completed; by which time the succeeding division (having received the word *Shoulders forward* on reaching the wheeling point) will have wheeled so many paces as the arc of the quarter circle exceeds the radius (or distances between divisions) being in the proportion of 11 to 7: Then the pivot file of the succeeding division, moving round the point, will occupy the identical spot from which the pivot of the preceding division had moved *forward*: Thus the wheeling angle is in succession left clear, and division after division leads out in the new direction at regular distances.

Wheel of divisions may be made forward or backward.

5. Wheels of divisions may be made either forward or backward. In progressive movement they are made forward, but particular occasions require that they should be made BACKWARD, on the pivot flank. In this manner the line may wheel into open column of companies, sub-divisions, or sections, and be prolonged when necessary to either flank, the pivots being thus preserved. The wheel backwards is also advantageously used in marching off parades, where guards are of different strengths, and is often essentially necessary in narrow grounds. By this means, although divisions should be unequal, either in the same battalion, or in a line, yet all their pivot flanks will after the wheel remain truly dressed; of course, the distances will be just, the line of marching accurately preserved, and each division, by afterwards wheeling up, will exactly occupy the identical ground it quitted: The backwards wheel, however, need not be practised where the ground is uneven, and the divi-

sions stronger than 15 or 16 file: Where this is the case, the divisions may FACE ABOUT, WHEEL, and then *Halt*, *Front*.

How performed.

6. In wheeling BACKWARD from line into column, when the right is to be in front, the wheel is made on the left; and when the left is to be in front, the wheel is made on the right. In wheeling FORWARD the standing flank-man faces outward from his division: In wheeling BACKWARD, he faces inward to his division. In wheeling FORWARD, the proper pivot flank of the column is the wheeling one: In wheeling BACKWARD, the pivot flank is the standing one, and remaining fixed, the divisions, however unequal, will always cover on that hand, which will not be the case if the wheel is made forward. In wheeling FORWARD, the command is TO THE RIGHT, (or) TO THE LEFT WHEEL: In wheeling BACKWARD, the command is ON THE RIGHT, (or) ON THE LEFT BACKWARD WHEEL; but when divisions are strong, or the ground uneven, the command will be FORM OPEN COLUMN RIGHT (or left) IN FRONT; upon which the pivot men face as required, and the divisions are FACED TO THE RIGHT ABOUT—RIGHT (or left) WHEEL; and are *halted* and *fronted* by the leaders of each when the wheel is performed.

7. If the divisions of a battalion are kept equalized (which they ought always to be for the purposes of manœuvre), they may wheel in succession upon the reverse as well as the pivot flank, without deranging the line of covering: But where divisions happen to be unequal, and that they are to wheel successively on the reverse flank, a strong division must wheel at a point

a point short of the preceding one, by the space of as many files as the strength of the latter is exceeded; and a weak division must overpass the wheeling point of its preceding stronger one, by the space of as many files as it is deficient.

Necessary recollections. Pl. 1. Fig. A. Fig. F.

8. It appears that the number of paces of 30 inches comprised in the front of any division or body, is, nearly 3-4ths of the number of files of which it is composed.—That the number of wheeling paces of 30 inches contained in the arc of the quarter circle, described by any division, is equal to the number of files composing its front.—That the number of files being once ascertained in each division, the officer commanding it must on all occasions recollect the number of paces that are equal to his front; also the number of wheeling paces which the flank man must take to complete the quarter circle; also the spare time, which he has to regulate the *Halt, March,* of his division after wheeling on the halted pivot.

TABLE OF THE NUMBER OF PACES CORRESPONDING TO A GIVEN NUMBER OF FILES.

Number of Files in a division, each occupying 21 inches.	5	10	12	14	15	16	18	20	30	40	50	100
Front of divisions in paces of 30 inches.	Paces In. 3 15	7	8 12	9 24	10 15	11 6	12 18	14	21	28	35	70

(For any number of files under 36, see also the scale of paces over Fig. A, Plate I.)

The

Fig. F. Wheeling paces required to describe

The 6th of the circle or an angle of	60°	are	$\frac{2}{3}$
The 8th	45°	—	$\frac{1}{2}$
The 16th	$22\frac{1}{2}$°	—	$\frac{1}{4}$
The 32d	$11\frac{1}{4}$°	—	$\frac{1}{8}$

of the number of files of which the front consists.

11. The field-officers and adjutants must always recollect the number of paces the front of the battalion and its divisions occupy, in order to take up ground exactly in all formations.

S. 72. *Movements.*

1. Every movement must be divided into its distinct parts, and each part executed by its explanatory and separate words of command.

2. Alterations of position in considerable bodies should begin from a previous halt; except where a new direction is given to the heads of columns, which may be done while in motion, or where their front is increased or diminished.

3. The exercise of small bodies, when within the command of one voice, appears more showy from the effect produced by keeping such bodies constantly in motion, and by changing from one manœuvre to another while on the march.—But such movements, and the formations made from them, must be on accidental points, unless mounted officers are very alert; and however brilliant and effective in battalion practice, they can only be considered as occasional exceptions,

not

not applicable to large bodies, where hurry must be avoided, and where concert and relative position are indispensable. It is at the same time most essential that Commanding Officers should not linger between their movements; to avoid which the caution for the succeeding movement may be given previous to the halt, which will afford the Field Officers and Adjutants time to direct the points for formation.—In moving to form upon any part of a battalion or brigade, the part or parts to be formed upon, may alone be halted—the other divisions continuing the march.

4. Movements and firings may be performed with unfixed bayonets, as giving greater ease to the men, and more accuracy of aim. They will be fixed upon all occasions when the approach of cavalry is expected, or when a charge is either to be made or repelled.

S. 73. *The Alignement.*

1. To march in an ALIGNEMENT is to make troops march in any straight line which joins two given points, —or to form upon any such given line.

2. In formations of defence the lines occupied may be curved, and following the advantages of the ground, but in those of attack, the lines must be straight, otherwise the troops in advancing must inevitably fall into confusion.

3. When troops are to form in a straight line, two necessary points in it must always be previously ascer-
Fig. 8. tained. One the point of APPUI (A. *a. a.*) at which one

one flank of the body, whether small or great, is to be placed, and the other the point of FORMATION or DRESSING (D.) on which the front of the body is directed.

Fig. 9. 4. When battalions, or divisions of a battalion, come up successively into line, the outward flank of the last formed and halted body is always considered as the point of APPUI (*a.*) or support of the succeeding one, (*d*).—The look and alignement of the soldier in forming is always towards the point of Appui (*a*), and the correction of dressing is always from that point towards the opposite hand D.—This great principle is to be observed, from the smallest body to the most considerable corps.

Attention required from the leading officer.

5. The officer who commands the division which leads a column in any given alignement must move upon an object without regarding his division, so that his shoulder shall just graze the head of any mounted officer's horse posted at an intermediate point, (or the shoulder of any man on foot placed for the same purpose), and which he must invariably preserve in a straight line with the distant points. All the following officers must maintain their exact distance from the company preceding, covering the pivots in their own persons. Should any of the companies neglect their covering, those that succeed them must rectify the fault and exactly touch whatever intermediate points may be placed to mark an alignement.

Wheeling into a new alignement

6. The leading officer must never change the time or length of step. He must move in one constant position with his front rank perpendicular to the line on which

which he marches. The same directions apply to the other officers who conduct companies; and they must also carefully observe, in wheeling, either on a halted or a moveable pivot, that no time is lost in giving the word *March* or *Forward*, immediately after the wheel, without attending to the succeeding company; which, if the officers preserve their due distances, and the wheels are made at the identical point where the leading company has wheeled, cannot fail of following out in due time, as already set forth in No. 4, of S. 71.

Fig. 5.

7. In order to accustom young officers to preserve their distance whilst marching upon the alignement, they may occasionally march in slow time; but the practicability of marching in quick time upon an alignement with correct distances, is too well established to be any longer doubted, and so essential an acquirement, that too much pains cannot be taken to ensure accuracy and perfection.

8. The explanation of marching in an alignement is of course closely connected with the movements of column, and must be considered together. (Vide S. 76.)

9. In whatever manner the leading division of a battalion arrives in a straight alignement in which it is to form, a mounted officer gives the point where it enters; and when arrived on the ground, and halted previous to formation, a field officer corrects, if necessary, from that division, the pivot files on the fixed distant points before the line is formed.

S. 74. *Points of Formation.*

1. When a single battalion takes up a new position, it may not seem material whether a flank is accurately placed upon a given point, or whether a line is exactly formed in any certain direction. But when it forms a part of a more considerable body, all its positions are relative to other battalions; and if the formations on flank points are not accurate and just, false directions and distances will be given to a whole line. It is apparent, therefore, that every single battalion should be accustomed to make its changes or position and formations on its own fixed flank-points; and commanding officers should always hold this object in view, and instruct their adjutants and others accordingly.

Relative connexions of battalion points with general formations.

2. The line on which troops move, or are successively to form, may be taken up to any extent by the prolongation of an original short base, given in the direction which the commander of a line will point out, according to his own views. This given base will serve to correct the position of the different persons who successively, as their separate corps or divisions require it, prolong the alignement from the several points already established. In general, therefore, the point (A), where a formation, or entry, into an alignement is to be made, being marked by a fixed person, the commander will place a second (*o*) thirty or forty paces without the first, exactly in the direction which he determines to give to his new alignement. These two persons will mark a base, which, by adjutants (*a.a.a.*) or others successively aligning themselves

Fig. 8.

themselves on the two first placed points and on each other, may be prolonged to any required length, at the same time that the commander fixes for himself, if the line extends beyond one battalion, a distant point (D) upon which to correct them.

Prolongers of alignements. 3. When the persons who prolong a line are on horseback, the head of the horse of each standing perpendicular to that line is the object; and when they dismount, (which should always be done, when the steadiness of the horse cannot be depended upon) they should cover, holding the horse by the outer hand, and the inner shoulder which is in a line with the horse will be rased in passing by the flanks of a column in march. All other men, who may be posted on foot at points for a column to march upon, will be placed in a similar manner. Although the leader of the first division of a column will march on the persons placed in the line, yet if the direction happens to be on some remarkable object, it should be fixed upon as the general point of correction. When columns move in prolongation to the front, two points (*o a*) will be given in front of the pivot flank of the leading division; and when they move by the flank march of divisions, two points (*o a*) will also be given for the flank march of the regulating division. Fig. 10.

Battalion points. Fig. 8. 4. Where two points (*o.* A.) are to be given in a certain direction, towards an outer flank (D.), the inner one (A.) should be first determined, and the outer one (*o.*) is immediately and easily taken over the inner point, and the flank one (D.) of correction. Should the outer one (*o*) be first taken, time is lost in directing the shifting of the inner one (A.), before it is truly aligned

aligned on the point (D.); besides the point (A.) in many changes of the position of a line or column is naturally the first ascertained, (being the pivot flank of a company on which the change is to be made, or the point of march towards which the column is moving), and from thence the flank point (D.) is then taken, which gives the new direction, according to the intention of the commander, who will align it with the distant point or object taken by himself; and the easy determination of (*o*.) follows of course.

5. In the successive formation of divisions into line, as from close column, from echellon, &c., the division

Fig. 9. upon which the formation is ordered (A. *a*.) will be considered the established base, which is successively prolonged for the others; and the divisions, as they come up, must align themselves correctly on the part already formed, and the field officer from the flanks of the base will correct that line in the true prolongation which is prepared for him by the adjutant, or other persons, just beyond where the flank of his battalion is to extend (D); and thus battalion after battalion arrives in line. When such formations are on a central division, these points of correction will be taken on both flanks.

Intermediate points

6. But however essential these battalion points may be, for the prolongation of extensive lines, they cannot be depended upon for correct formations. Their view may be obscured or intercepted by intervening obstacles, and officers commanding companies require more certain and intermediate guides when forming from a point of appui. It is material that these aids should be afforded, for the mistake of any one officer, who fails

fails from any cause whatever, in the regular formation of hisdivision upon the flank point of the battalion, must undoubtedly derange the succeeding divisions, if they are dependant upon the correctness of those in their front.

Covering serjeants. 7. To remedy the inconvenience, therefore, which has been found to attend the formation upon distant or battalion points alone, two serjeants will place themselves upon *all occasions* of formation of lines, in front of each flank file of the company named as the base, facing to the point of appui (at A *a*): and a covering serjeant will run out from each company as it arrives within eight paces of the point of formation, and will cover (*c. c. c.*) at the distance of the division in the line established by the serjeants in front of the base, (A. *a.*) who will remain steady until the third company from them receives the words "*eyes front,*" when they will pass to the rear. The covering serjeant of the next company will also remain steady until the third company from him receives the words "*eyes front,*" and so on until the line is formed. In this manner each covering serjeant will have two points to cover upon; and each division will always have two points upon which it will be brought parallel to the general line, and independent of the divisions already formed; and if the points are well taken up, and the divisions properly conducted, little dressing will be necessary. When covering serjeants are taking up points, they must hold their pikes perpendicular in front of their bodies.

Fig. 9.

Points to be always kept clear. 8. It is most essential that these serjeants, who are so many intermediate points, connecting the flank of appui

appui with the outer flank point of the battalion, should be always kept clear for the view of division officers. Divisions must therefore "*dress up*" to them *upon all occasions*. (Vide *Deployments and Echellon Movements*, S. 113, S. 117, and S. 118.)

Base points.

9. Whenever a company is named as the base of a formation, two serjeants belonging to it will immediately spring out and give the base for the covering points. As formations will generally be made upon the flanks or centre, two serjeants of each flank company, and of the two centre companies, will always be told off for this purpose. When formations are made upon the centre, the intermediate points of each wing will cover the central base; and upon change of front on the colours, one of the supernumerary serfeants of the centre will place himself in the new front before the colours, and the covering serjeants of the two centre companies will change places and cover him, the coverers of each wing aligning on them as before.

Covering serjeants in changes of position by the open column.

10. When a line changes its front in any direction, by means of the open column, a covering serjeant from each division will always run out ten paces before the division reaches the new alignement to mark its distance; and he will cover at the proper pivot flank upon the point of formation.

11. When formations to line or changes of position are made by subdivisions or sections, the covering serjeants of companies are sufficient to take up points.

Corrections upon flank pivots.

12. The covering of serjeants, whether in line or column, will always be corrected by the field officers from

from the flanks, or from the base of formation, upon the flank point of the battalion.

Adjutants and their aids.

13. It is the particular duty of an adjutant in the field, under the superintendance of the field officers, to establish the points necessary for the movements and formations of the battalion, and he may be assisted in the exercise of it by two detached persons placed behind each flank of the battalion, who are properly trained, to take up quickly such line as he he shall give them. These aids may be employed to give a succession of points on which a column may march: When the head of the column shall reach one of them he will move 30 or 40 yards beyond the other aid, and will cover under the direction of the adjutant, so that the column will thus always have a succession of points to mark the alignement. These aids should carry small flags attached to a staff about two feet long.

14. All mounted officers, whether staff or regimental, ought to have a good seat and a good hand on horseback Their horses should be well managed, so as to be thrown into a full canter at once, to take up ground, either for the depth of a corps in column, or for its extent in line: And every officer should accurately possess himself of the rate at which his horse may canter, so as to calculate how many strokes of a full canter bear a proportion to the space occupied by files of infantry at 21 inches from each other. The rate at which horses canter is so variable, as to render it impossible to establish any accurate table of calculation on this head; and therefore it must rest with each individual officer to ascertain the paces of his horse,

horse, and thus to take up distances with preci sion.

S. 75. *Dressing.*

1. All DRESSING is to be made with as much alacrity as possible, and the dresser of each body, as he accomplishes the operation, will give the word, *Eyes Front*, that heads may be replaced square to the front. If the body to be dressed is extensive, as that of a battalion or parade, the dresser must place one division justly before he proceeds on that which is beyond it: And great care must be taken that the soldiers shall wait for the successive movements of the files nearest to the point of *Appui*; so that the dressing shall be taken up gradually from thence to the outer flank, which by this means will avoid the error of passing the line.

2. When the leaders of companies dress their divisions, the duty must be performed without noise or bawling to the men, which has an unsteady and unmilitary appearance. If it be necessary to call any particular men to dress up or back, the caution must be given in such a low tone as to be heard only by the company.

Fig.. 9.

General attentions of dressing in all formations.

3. When FORMING into and DRESSING in line from column, and also forming line from echellon, the soldiers come into line with their eyes directed to the general point *Appui* (A.), where the leading flank is to rest: The officer in dressing is placed on that flank of his division, to which the men's eyes are turned, and

and from the second file from the flank of the company towards which his wheeling flank moves from column, or his inward flank from echellon, he makes his corrections on his intermediate point, and the battalion point (D.), which is previously marked by the adjutant, or some other person placed in the true general line, (vide S. 74). On all occasions, therefore, by the men aligning themselves to one hand, and the officers correcting to the other, the most perfect line may be obtained.

Fig. 9.

4. When so many points are given, as detailed in the last section, it becomes easy to dress correctly a company or battalion after wheeling up, if due care be taken that the pivot men do on no account move up, or fall back, whatever direction may be given by the company officers for completing the dressing. If a defect exists after a wheel into line, it must proceed from the other men not having aligned with those fixed points; the internal correction of companies must therefore be made, but the original pivot men remain immoveable, until a general dressing of the battalion is made by a field officer, if necessary.

Closing to correct distances.

5. When distances have been lost, and that it is necessary to close by the side step to regain them, such closing will always be made to the centre, which is the point of *appui*: The dressing will therefore be made from centre to flanks; and officers commanding companies of the right wing will place themselves, to give the word and time, in front of their left flanks, and those of the left wing in front of the right flanks. When the whole battalion is to close, the flank to which the closing is made, will be the point of *appui* from

rom whence dressing is taken; and officers will accordingly place themselves in front of the flank of their companies to which the closing is ordered. When officers change from one flank to the other in order to close, they will pass by the front, instead of the rear of their companies.

Dressing a battalion after an advance in line.

6. In *dressing* a single battalion after the halt, whatever correction is necessary must be made, by advancing or retiring the flanks, and not by moving the centre; which has been the guide during the march. When the commanding officer gives the word Dress, the company officer on the left of the colours instantly dresses the six or eight files to the right of the colour in a proper parallel direction, the two wings immediately conform to the centre, and afterwards receive the word Eyes Front. Should the commander require a more exact dressing, he will order a colour to advance one step, and face to the left, also the second company officer on the left of the colour to advance one step, and face to the right; then the flank company officers to advance, and to face to the centre; then each other company officer instantly to cover those at their due distances, and face to the centre; then the officers of the left wing to face about, so as the whole stand fronted to the left.—Then battalion, right dress, on which the companies march up to their respective officers, who will halt and dress their companies, and immediately front into line.

7. It must be observed in this mode of dressing, whether it is taken from the centre, or from a flank, that company officers, who originally face to the left,

take distances equal to the front of their own company, from the officer before them; but such as face to the right, must take distances from the officer before them equal to the front of the company, which in line is on the right of them. When circumstances allow the dressing to begin from the left, an advantage arises that the officers do all originally face to the left.

8. In all dressing, the rear rank men must conform to the movements of their relative files of the front rank while dressing from a point of *Appui*; and the supernumerary officers must be responsible that the dressing of the rear rank is thus accurately preserved.

S. 76. *Open Column.*

Formation of columns

1. Columns are formed from line for the convenience of movement, and for the purpose of again extending into line. Every column of march or manœuvre must be formed by a regular succession of the divisions from right to left of the line, or of such of its parts as compose the column; for whatever is the relative position of a body in line, such ought it to be in column; and where several connected columns are formed, the same flanks of each should be in front; but whether the right or left, will depend on circumstances.

Formation of open column from line.

2. When a column is to be formed from line, it may be done by the wheel of divisions, either backward or forward, as already described in S. 71. Upon the caution, the leaders of divisions place themselves close before the centre of their companies, facing to the front;

front; the pivots face, and the covering serjeant of the right, or left, company (according to which flank is to be in front) runs out and places himself at the point where the wheeling flank of that company is to rest at the completion of the wheel; the covering serjeants of the whole fall back two paces; and the supernumerary rank closes up within two paces of the rear rank. When the wheel is performed, as directed in S. 49, the officer corrects the dressing, and places himself on the pivot flank: his covering serjeant covers the second file from that flank.

Column of march and manœuvre

3. Columns of march or manœuvre will generally be composed of companies, sub-divisions, or sections. An open column occupies the same extent of ground as when in line, minus the front of its leading division; and its chief objects are facility of movement, the quick formation of the line to the flank, and the change of situation in the shortest lines from one position to another.—It is named the column of march or ROUTE, when applied to common marches, where the attention of men and officers is less scrupulously demanded—and the column of MANŒUVRE, when being within reach of the enemy, or at exercise, the greatest exactness is required to ensure its speedy formation at any instant into line, during its movement from one position to another.

Central movements in double columns.

4. Double columns are formed upon the centre of battalions, brigades, or lines, for the special purposes of attack, and, in certain cases, for the passage of defilés when presented in front of a centre. (Vide S. 86.)—All such advances from the centre should be in double column, formed on the two centre sub-

divisions of a battalion, or the two centre divisions of a brigade. It is obvious, that, for the purposes of attack, a force can, by means of this formation, be more readily and sooner brought to bear upon a given object, than by advance of column from a flank:—In the latter movement the divisions have the diameter, in the former only the radius, of the circle to traverse:—It follows, therefore, that the same force is formed from the centre in one-half the time that it can be done from a flank:—But, in battalion movements, the consideration of these advantages must not prevent the equal practice of advances in column from a flank; for when the central movement is applied to a brigade, the formation from the centre for the whole becomes totally a flank movement for the battalions on the flanks. Thus battalions should equally practise flank movements in column to the front or rear, in order to perform with readiness the part that may be required of them in the central movements of larger bodies.

Application of central columns, and columns from a flank.

5. For the passage of defilés, the movement in single column from a flank is preferable to that in double column from the centre, particularly where the defilé happens so to present itself in front, as to be unfavourable to a direct debouchè from the centre: Where this is the case, a battalion, or brigade, may be conducted to it with much greater facility in single column from a flank; and the front of the column, in the latter movement, can always be diminished to the smallest degree of reduction, according to the width of the defilé; whereas objections may be offered to the general application of a debouché in double column from the centre *for the passage of defilés*, inasmuch as it

Fig. 19

it is not susceptible of a similar reduction of front without being liable to the confusion which may arise in narrow and difficult passes, from the probable mixture of files, if diminished to less than double sections: But where the debouché can be made in a direct line from the centre, and where the defilé affords room for the double column formation, these objections do not exist. For the purposes of attack, however, the advance in double column is at all times to be preferred, upon the grounds explained in the last paragraph; and a column which advances from a flank, may, after passing a defilé, easily assume this formation,—if in brigade, by the front half of the column forming, to the proper flank, abreast of the rear half;—if in battalion, as instructed in S. 102. It may further be observed, that exclusively of the advantages derived from the central double column for objects of attack, this mode of advance is susceptible of the most ready reformation of the line to either flank (Vide S. 86. No. 5.) or to the rear upon the front, or rear, divisions (Vide S. 86. No. 4 & 6). With a view to combine, therefore, as far as possible, the separate advantages of the flank and central movements, it may be laid down as a principle, that the passage of a defilé should generally be effected by the march of a single column from a flank, and that advances for objects of attack, where the front is open, should be conducted in double column from the centre.

Changes of position, or the front of lines, by the open column.

6. Changes in the position, or the front of lines, are generally made by means of the open column, where the difficulty of the ground does not favour a movement in echellon; or where the necessity for the latter march is not called for by the immediate

presence

presence of an enemy: such changes of position are effected to the front or rear of a flank division, or upon the centre division of a battalion or brigade formed in line—by all the remaining divisions facing by threes, and disengaging their heads towards the points of formation—and proceeding by the flank march of threes to their respective distances in open column, as taken up by their covering serjeants. When a whole battalion moves by such flank marches of divisions in open column, the commander must be careful to keep the heads of divisions as square as possible with the Fig. 34. conducting division, in order to be ready for any formation that may be necessary to resist attack on the march; for this purpose, the leading division will occasionally mark time, and the other divisions will move up at the *Quick* or *Double March* as may be necessary.

7. In all changes of position on a central or flank division, by means of the open column, the parallelism of the divisions, and their true perpendicular position with the intended alignment, are more readily attained by forming a column, facing towards such named divisions (Vide S. 93, 94). The whole of the divisions, therefore, whether the change is upon a central, or a flank division, will face to the right or left according to which flank is to be in front, and, those in front of the formed ones, will countermarch and move to the proper pivot in column—double distance being taken by the division in front of the formed one, so that the double column may wheel to the right and left into line. Should it be intended to move previous to wheeling into line, double distance need not be taken, and in that case, the divisions in the flank towards which

which the movement is to be made, will countermarch, or prolong the alignement for a short distance with the rear rank in front.

8. When an open column is moving to a flank by threes, it is most essential that the commander should bear in mind, (for it cannot be readily ascertained by the eye,) which flank of the column would be in front if it were halted and fronted. He will therefore impress upon his recollection, that when divisions lead out from line to the rear from their right flanks, *the right of the column is in front*, and the word is THREES RIGHT, RIGHT WHEEL; and when to the rear from their left flanks, *the left of the column is in front*, and the word is THREES LEFT, LEFT WHEEL. When they lead out to the front from their right flanks, *the left of the column is in front*, and the word is THREES RIGHT, LEFT WHEEL; and when to the front from their left flanks, *the right of the column is in front*, and the word is THREES LEFT, RIGHT WHEEL. An open column may effect a change of position upon its front, rear, or any central division, by the named division, wheeling up according to the front to which it is intended to change; and the other divisions facing, and proceeding as before.

Changes of position in column to the front, on the fixed or moveable pivots.

9. In changes of position, by means of the open column from a flank to the front, the whole line, except the leading division, will previously wheel backwards or forwards by divisions into column: If the wheel is made backward, the flank division may stand fast until the column has marched up to it; and it then proceeds by command of its own leader; and the succeeding divisions will follow in column on the

the fixed or moveable pivot: But changes from line to the front, in open column, may be made by the flank division moving at the same time that the remaining divisions are thrown into column by the inclination of the outer shoulders, and following the advanced division on the moveable pivot, by which considerable time will be saved. Vide S. 86.

10. There are occasions when the change of position is made on the moveable pivot, and yet a previous wheel into column on a fixed pivot may be necessary; as in the retreat of a line in double column from the centre (Vide S. 87. No. 2), where the wings must wheel backwards upon different flanks (forming two columns facing each other), which leaves room for the wheel and retreat of the two centre divisions.

11. Also when a battalion is to march off by divisions from a flank to the rear, it is done with more accuracy by previously wheeling into column on a fixed pivot, and then wheeling in succession to the rear on the moveable pivot as before.

Covering of pivots.

12. In column, divisions cover and dress to the proper pivot flank; to the left when the right is in front; and to the right when the left is in front.—The *proper* pivot flank in column is that which, when wheeled up to, preserves the divisions of the line in their natural order, and to their proper front; the other is called the *reverse* flank.

Posting of officers.

13. In column, the ordinary post of a commanding officer of a battalion is near the flank of the leading division; that of the other field officers to be respectively

tively near the flanks of the centre and rear on the reverse flank.

In column of companies. 14. Each division of which a column is composed is conducted by a leader, placed on the pivot flank of the front rank, which is his general post: In a column of companies such leader is the captain or commander of the company. The lieutenant is to cover the second file from the reverse flank; the second and third subalterns in rear of the second and third sections; and the covering serjeant is to cover the second file from the pivot flank.

In column of sub-divisions. 15. In a column of sub-divisions the Captain leads the head sub-division of his company, and his Lieutenant the second. The second subaltern is to cover the second file from the reverse flank of the head sub-division, and if there is a third subaltern, he will cover the second file from the reverse flank of the second sub-division.

In column of sections and threes. 16. In a column of sections, the Captain leads the head section of his company; his lieutenant the rear one; the second subaltern the third one; and when there are four sections, if there is not a third subaltern, the second section will be led by the covering serjeant.—When the establishment admits of a third subaltern, he will command the second section; and in that case the covering serjeant will cover the second file from the pivot flank of the leading section, and he remaining serjeants will form in rear of the other sections.—In the march by *threes*, the captain is placed on the inner flank of the leading section of threes—his covering serjeant leads the centre file—and the other

other officers and serjeants are formed in the rear, or reverse flank, in the same order as when in line.

17. When any considerable continuation of the march is the object, and that pivot officers are permitted to be in front of their divisions, their flank posts must be occupied by non-commissioned officers, who remain answerable under their direction for the preservation of the proper distances.

Music, drummers, &c. 18. In open column, music, drummers, &c., of battalions, wheel with and remain closed up to the rear of their respective divisions.—In column at half or quarter distance, they may occasionally, if there is space, move by threes on the reverse flank.—Instead of being kept collected, they may in column of march be sent to their respective companies to remain in the rear of each; but they are on no occasion to lengthen out, or interfere with the movements in column, or to increase the intervals between battalions in column.

Marching in column; by threes —and by files. 19. It has been already observed, that all marches of battalions are made in column of companies, of sub-divisions, or of sections: But a battalion, or brigade, in route marching, may be formed by *threes*, particularly where the roads do not admit a column of more extended front. The march *by threes* may also be applied to the internal movements and formations of the divisions of a battalion, and to the flank movements of battalions in columns of manœuvre.—This formation and march afford the advantage of moving the divisions of a battalion in compact order towards their several points of formation, upon a space within the extent of their front; and the *double time* can always

Fig. B.

always be applied, without disturbing that compact Fig. B. order of march: whereas *file marching* is always liable to an extension of the files, *much* beyond the space which the front of divisions would occupy; and the *double time* could never be applied, without causing an additional lengthening out of the rear, which must create disorder, and occasion the loss of men in action. It may be observed, that the formation *by threes* is liable to the repeated derangement which the casualties in action may cause in the telling off; but this probable inconvenience appears to be counterbalanced, by the advantages of this compact order of march, as well as by the facility it affords of moving in double time; and it is the duty of the officers and non-commissioned officers in the supernumerary rank, to see that the breaks occasioned in the telling off by threes, by casualties, may be corrected as far as possible, previous to a flank movement; and to renew the general telling off from the right of a division, when it can be conveniently accomplished. Until these corrections are effected, each file will preserve and form according to its original telling, notwithstanding the casualties around it; and in cases of emergency, where such breaks cause a total derangement of the *threes*, and where an immediate renewal of the telling is impracticable, it is always in the power of officers commanding divisions to resort to file marching, (for which the drill is preserved complete, vide Part II.) until the telling off can be re-arranged.—*File marching*, in its general application, therefore, is only to be used in the following instances; *viz.*,—1st, In changes of position in column, where a wood, or the nature of the ground, may render it difficult to preserve the formation of *threes*: 2dly, In the countermarch of companies

companies in column: 3rdly, Where a division is to gain ground to a flank to an extent not more than double the space of its own front: 4thly. In passing a defilé where the width will not admit any column formation: 5thly, In the march of small guards, or parties, through streets or other situations, where column marching is inconvenient: And lastly, Where a division, having had its telling off by *threes* deranged, cannot be conveniently re-told, previous to a flank movement.

20. It must be recollected, that in the formation of threes, the third file doubles up behind the front rank man of the second file; therefore, in moving to a pivot on the left when forming column, &c. &c., the leaders of companies must be instructed to halt their divisions when No. 2 file shall be one full pace from the pivot, in order to leave room for No. 3 file to form up on the word, *Halt, front*. The same attention will be required when moving to the right, in order to form on a left pivot; in this case, No. 2 file must be allowed to pass the pivot one full pace, to leave room for No. 3 to form up.

Application of the four deep formation.

21. The formation of four deep, as instructed in S. 47, may be applied with great advantage to the passage of lines through each other; and for bringing the rear of a column (formed at any but close distance) to the front, where there is no room for a counter-
Fig. C. march along the flanks (vide S. 111. No. 2.) Four
Fig. 44. deep may also be applied in certain cases, to the flank movements of divisions, or battalions, upon a space within the front occupied when formed in line; and in which its single rank formation, affords room for moving

moving without restraint at the double march; as well as with trailed arms when moving to the ground of exercise, or upon any other occasion, with unfixed bayonets. Four deep may also be applied to the rapid formation of square from line, (Vide S. 89. No. 5.) and occasionally to the change of front of lines by the flank march of divisions in open column.

22. When divisions move four deep to the right or left, it must be recollected, that in both cases there is a file to form up on the outer flanks, upon the *Halt, front*. When forming on a pivot point, therefore, the leaders of companies must leave room for this file, as directed for the threes, in No. 20.

Opening or closing of rear ranks.

23. When the rear ranks close or open on the march, in the one case they will step nimbly up, in the other they will slacken their pace, until the due distance is attained. In both cases the front rank continues to proceed at its original rate of march.

Counter-march of double column.

24. The counter-march of the double column from both flanks, in which the divisions change places at the same time, may be made by threes, and it prepares such a column, to any extent, for formation of line to its former rear (Vide S. 56, No. 3). The change of places of divisions of a double column without a countermarch, either inverts, or relieves a double column from inversion. (Vide S. 87, No. 3.)

Counter-march by files.

25. The countermarch of the divisions of a column, each on its own ground, will generally be made by files and it changes a column that is standing with its right in front, into a column with its left in front, and thereby

enabels

enables it to return along the ground it has gone over, and to take new positions without altering or inverting the proper front of the line. The countermarch by ranks has the same effect, and is more expeditious; it is peculiarly adapted to the close column where there is no room for the countermarch by files: All countermarches of the divisions of a column on their own ground will be made at the quick, and never at the double march.

Counter-march by divisions. Fig. 39.—Fig. 40, 44.

26. The countermarch by divisions successively from the rear to the front, changes the leading flank of the battalion column, but allows it to continue its former direction of march, and is a previous manœuvre often necessary to enable a battalion to take up a relative position, or to form in a mass of columns.

27. All countermarches necessarily change the pivot flanks of columns. The colours cover the third files of men from the pivot, and must be ready to move up when the line is to be formed.

Formation of line from open column. Fig. 32.

28. When the column of companies halts to form, pivot flanks are in an instant corrected from the leading division by field officers of battalions. On the caution, officers move to the centre of their companies three paces from the front rank; their covering serjeants place themselves on the right (g) of each if the wheel is to be to the left; or behind the pivot file if the wheel is to be to the right, and a serjeant of the leading company of the battalion runs up and places himself in the new alignement to mark where the wheeling flank of that company is to rest. Pivot men of the front rank, face square into the new direc-

tion

tion and the rear rank locks up to the front rank. The whole wheel up and halt.—Officers dress the interior of their companies, and then replace their serjeants, who are now in the front rank.—If any farther dressing is necessary, it must be ordered and made by a mounted field officer.

29. In general the whole of a battalion will be halted on its ground, stand in column, and have its pivots adjusted, before it wheels up and forms: but if necessary, and where parts of it arrive in the line by threes, they may form successively as they come up. If part of a battalion should therefore be ordered to wheel into line while the other divisions are not yet in it, the pivot men of those divisions (and not the officers) must cover on the formed part of the line before they wheel up. And when several battalion columns changing position enter separately, and are to form in the same line, each may be successively wheeled up, if so ordered or intended, when its adjoining one has three or four of its divisions standing in column on the line.

Disengaging heads of companies or divisions.

30. In many cases as already explained, independent of the passage of lines, formation of close columns, &c., the companies or divisions of a battalion in line are ordered to face by threes, and to lead out to the front or rear, into column, without first wheeling the quarter circle. An explanatory caution being given, the companies face to the point directed, and at the same instant the two leading threes of each throw themselves to the flank, according as they are to move, so as to be disengaged from the flank of the preceding company.—In this situation each leader is enabled

enabled, at the word QUICK MARCH, to move independently, without check, preserving his distance from whatever is the proper front of the column.

Column of march.

31. The rear divisions of a battalion, or more considerable column, either of march or manœuvre, will constantly follow every turning which the head may make; each successively changing its direction at the same point with the leading division; and, although in route marching the files of a division may be permitted to loosen, and move with freedom, yet the pivot files must carefully preserve their place and distances *at all times,* under the constant superintendence of an officer of each division: In long marches officers and pivot files may be frequently relieved in these essential attentions: Preparatory to any relaxation in route marching, the words MARCH AT EASE should be given, when the soldiers may be allowed to open their files, carry their arms as they please, and converse. The officers may march likewise at ease, and, with the exception of the officer charged with the superintendence of the pivot files, they will be found most useful, in rear of their companies. At the word ATTENTION, files are closed, the step taken up, silence preserved, and arms sloped. In this restored order all alterations of front, formations, &c., should be executed. When it is probable that a column of march may be required to form to a flank, it will move at open distance, and when ordered to HALT, and FORM in line, the pivot men of companies must remain steady where they are found at the word HALT; and the companies or divisions will wheel up into what will probably be a curved but a just line.—When the march is again to be resumed, the line breaks

breaks into column, and the rear divisions at their ordered distances will continue to follow the exact path traced out by the leading flank: the following divisions of a column must never deviate from this rule, or endeavour of themselves to get into a straight line, when the general direction is a winding one, until an express order is given for that purpose; which can hardly ever be the case until the head of the column is halted, with a determination to form the line in a straight direction. When the necessity of forming to a flank is not likely to occur, the column of march will be at quarter distance (Vide S. 77, No. 3.): But the above attentions from the pivot files and division officers are still essential.

Fig. 2. Diminishing or Increasing the front of a column.

32. The column of march or manœuvre, in consequence of obstructions in its route which it cannot surmount, is frequently obliged to diminish its front, and again to increase it when such difficulties are passed; this is a most important movement; and if a battalion cannot perform it with precision and attention, so as not to lengthen out in the smallest degree, it is not fit to move in the column of a considerable corps.

How performed.

33. The increase or diminution of the front of the column is performed by the battalion, when in movement, or when halted.—In movement this operation is either done by each company successively, when it arrives at the point where the leading one of the column performed it, or else by the whole companies of the battalion at the same moment by command of the commanding officer. In the latter case, at the instant the battalion should begin to reduce or increase

its

its front, the commander gives the general caution to that effect: if by companies successively, the commander of the battalion gives the caution to the leading division only, or if an obstacle presents itself unexpectedly, the commander of the leading division will proceed to diminish the front, without waiting for the direction of the commanding officer. The leaders of companies give their words of execution on reaching the obstacle, to the reverse sub-divisions or sections to double in front of the pivot sub-divisions, which preserve their original distances from each other, and never alter the pace at which the column was marching, but proceed as if they were totally unconnected with the operation that the others are performing. (Vide S. 59 and 60.)

Fig. 1—A. 34. When the front of a column of companies is to be diminished, it will always be done by the reverse sub-divisions, or sections, doubling to the pivot flank, so that the battalion may remain ready to form in line by a simple wheel up to that flank;—therefore the doubling will be in front of the left when the right is in front, or in front of the right when the
Fig. 1—B. left is front.—When the front of a column is to be increased, the reverse sub-divisions or sections will form on the pivot one, as instructed in Sect. 59.—This method of making the reverse flank divisions increase or diminish the front of a column, will appear to possess the advantage of leaving the pivot flanks undisturbed in their proper line of covering; and it may therefore be applied to a column of manœuvre, as well as of march, without deranging the general alignement. When the front of a double column is thus diminished, it must be recollected that the

the one has the right and the other the left in front; and that the doubling of the reverse divisions will leave between the two columns the interval which they occupied; the columns will therefore close from the *halt* by the side step, or, if in movement, by the diagonal march inwards; and they will open out in the same manner when the front is to be increased to give room for the reverse divisions.

March of the column through a wood, or in embarrassed ground. 35. The march in column through a thin wood, or in ground where impediments frequently change the direction of its head, or along the winding of heights which are to be occupied, will be best made by subdivisions, or by sections of five or six files in front.—

Fig. 11. The *pivot* files will preserve exact distances from each other, choose their own ground, and wind as the trees or other impediments permit, along a general direction:—When the column *halts* and forms, the line will be a continued curve, which can afterwards be easily corrected, if circumstances require it.—In such situations the *pivot* flank leaders should not double or quit the line of march if it can be avoided; but the other files may (if impediments are to be passed) open out from those pivots, who in the mean time preserve wheeling distances, and are in constant readiness to *halt* and form in line, the others closing in to them.

Obstacles in march of columns 36. Should the march of a column, whether in a straight alignement or otherwise, be at any time interrupted by pools of water, or any other obstacle which is impassable, the march will be continued direct to

Fig. 12—A that obstacle, (*b*) the obstacle will be surrounded (*c*) (and always if possible by deviating to the reverse flank so as to remain behind the line), and the same straight

 line

line will again on the other side be taken up by the pivots, at the point in it which a detached person has prepared (*o*).—Allowance will be made, when the line is to form, for the breadth of such obstacle, by dou-

Fig. 12—B bling as many divisions (*dd*) as will fill up the vacancy which has been so occasioned: Small interruptions must never, if possible, cause the pivots (*e*) to deviate from the straight line, when the intention of forming upon it is perfectly understood.

S. 77. *Column at Quarter Distance, and Close Column.*

Formation. 1. The column at quarter distance is formed either by closing from open distance, when it is intended to march upon the same alignement on which the open column had been formed, or by the march of companies or divisions to a flank, when it is intended, from line, to change the head of a column, and to form in rear of the right or left division of a battalion. This column is to be conducted upon an alignement, and upon all changes of direction, according to the principles which guide the movement of the open column; except that wheels and changes of direction in column at quarter distance must always be on a moveable pivot; distances in column equal to the front of divisions alone [admitting of halts and wheels upon a fixed pivot, without crowding and confusion at the

Distance of divisions in close columns angles. When close columns are formed, the companies or divisions must be at *two* paces' distance, in order to leave room for the formation of threes, measuring

suring from the heels of the rear rank to the heels of the front rank; and these distances must always be taken up by the covering serjeants; who will for that purpose run out a few paces before the leading flank of divisions shall reach the alignement: If the distance is to be taken up from the rear, he will cover the formed flank, and then go to the right about.

2. The column at half and quarter distance is equally applicable to most of the changes of position, and formations of open column, except the formation of line to a flank, or in changes of position where line is to be formed to a new front; and in all route marches, and marches of manœuvre to any considerable extent, the quarter distance should be adopted.

Advantages of quarter distance

3. Quarter distance unites the convenience of moving upon a space three-fourths less than the extent occupied by the march of an open column, with the capability of moving and forming in any manner (as detailed in the column evolutions) that may be required to resist attack: And the compact formation at quarter distance has, moreover, the advantage of averting the evils which attend the loss of distance in open column, from the irregularity of ground or other causes; for, even if intervals are lost on the march at quarter distance, the unity and strength of the column are not impaired, unless that loss exceeds the extent of the front of the division, which can rarely happen without marked and culpable inattention of the company or division leaders.

Applicable to all close column movements

4. The column at quarter distance also partakes of all the properties, and is capable of all the formations and deployments, applied to close column; and there

there may be few occasions where the latter is necessary, unless in the wheel of columns for change of front (which are performed in more compact order at close distance) for the assembly in mass of a number of troops in a small space, or to limit the space occupied by the march of a column. The wheel of a column at quarter distance may however be performed, as hereafter explained, under every circumstance, if deemed necessary.

Wheel of quarter distance or close column.

Fig. 49.

5. Much advantage attends the wheel of a column at quarter or close distance: it prepares a column for deployment on a change of front to a flank (*ac*) or to the rear (Vide S. 109 and S. 110); forms contiguous columns into a column in mass facing to a flank, and a mass into contiguous columns (Vide S. 139): it also facilitates the change of front of contiguous columns, in order to form a line to the rear without inversion (Vide S. 145); and changes the front of a close column by the wheel, and countermarch of subdivisions round the centre, (Vide S. 110.), so that a battalion column that is to take its place in a mass of contiguous columns, may thus change its wings on its own ground, to correspond with the front of the line of columns. Battalions must, therefore, be well practised in these wheels; and in order to perform them in close column, without crowding, at the inner flank, the officer or outward file, whether officer or man, must be the pivot upon which the wheel is made (otherwise officers and their covering serjeants must fall back as the pivot files face and cover), and thus the whole column wheels as a division under the direction of the commanding officer

Officers the pivots.

6. In

Rear divisions half face to the reverse flank.

Fig. 49.

6. In order also to avoid crowding in the wheel of close column, it is necessary that, upon the wheel being ordered, all the rear divisions make a half face to the reverse flank, each file, on the word *March*, and during the wheel, taking care to cover the one in front, by which the component parts of the column will be rendered flexible, and freedom will be given to the files to move in the course of their respective circles, without pressing inwards upon each other and upon the inward flank.

Special attentions to the wheel of a battalion at quarter distance.

7. In the same manner the rear divisions will make a half face to the reverse flank, when the wheel is made at quarter distance. But in order to preserve the quarter distance accurately, the flank file of the leading division on which the wheel is made, will upon the caution, advance six paces, halt, and face. The leading division will advance on the word *quick* or *double* march, and will wheel round this file at the usual pace, while each succeeding division will advance in circling round to quarter dis ance, which will leave room for the divisions of the rear wing to circle into their relative positions at that distance. Thus in all such wheels at quarter distance, the battalion must gain six paces to its front (Vide S. 109, No. 3.) In wheeling on the moveable pivot, the rear divisions make a *half turn* towards the shoulder brought forward, and the front division wheels and advances in the new direction, the rear divisions circling round. (Vide S. 109, No. 4.)

Posting of Officers.

8. In column at quarter distance, officers are psoted as laid down in the open column formation (Vide S. 76.) But in close column the supernumeraries will form

form on the reverse flanks of companies; and upon the caution for deployment, they will pass to the rear of the column, and will form from right to left, as their companies are numbered in the battalion: when the close column is to countermarch, they will remain on the reverse flank, and face with their companies: But instead of countermarching with them, they will countermarch on their own ground.

Deployment.

9. The column (close or at quarter distance) may form in the alignement of its front, rear, or central division, by the deployment or flank march by threes, by which it successively uncovers and extends its divisions into line; by their *Front and rear turn* towards the division on which the formation is made. (S. 113.) The supernumeraries will move with the deployment, and halt with their respective companies.

In alignement with the front division.

10. When a column deploys on a rear division, but in the alignement of the front division, the former, when uncovered, must move up to the identical spot

Fig. 55. which the front has quitted, and which its covering serjeants will mark (*ss*):—The points therefore necessary for the formation of the battalion will in this case be taken in prolongation of the front division, upon the flanks of which the base points will be placed (Vide S. 74, No. 9.); and the divisions which successively move up, must *Halt, front,* until their front is clear.

Attention to the deployment.

11. The flank march by threes, whether it is at the quick or double march, will be made firm and parallel to the general line. The field officers of the battalion will superintend its execution, and whenever a division is uncovered and fronted, or turned

to

to the front or rear, the covering serjeant will run out to mark the distance for the front of his division in the line, as directed in S. 74, No. 7., each leader bringing up and halting his division in the line, and dressing from the flank of appui on the serjeant and battalion point.—If the company is brought up at the *double march*, it will be halted two full paces in rear of the line, in order that it may neither press upon nor overshoot the points; and the officer, without unnecessary pause, will give the word *Dress up* from the flank of appui.

S. 78. *Echellon.*

Utility in changes of position.

1. The echellon position and movements are not only necessary and applicable to the immediate attacks and retreat of great bodies, but also to the oblique or direct changes of situation, which a battalion, or a more considerable corps already formed in line, may be obliged to make to the front or rear, or on a particular fixed division of the line.

How formed.

2. The oblique changes are produced by the wheel (less than the quarter circle) of divisions, which places them in the echellon situation. The direct changes are produced by the perpendicular and successive march of divisions from line to front or rear.

Fig. 13.

3. The march in line or in the direct echellon, B, produces new parallel positions to front or rear. The march in echellon, C, when formed by the wheel of the divisions from the line, produces new oblique positions to front or rear, according to the degree of wheel given to the echellon. The march in

open

open column, A, produces new prolonged positions to either flank.

Formed by the wheel of companies

4. The echellon of march in oblique changes of position, will be composed of companies or sub-divisions; and will generally be formed from line by the wheel of each on its own flank, to the hand to which it is to move. Such wheel will seldom exceed the eighth of the circle, but can never amount to the quarter circle, otherwise the corps would stand in open column.

Fig. 13.

Properties of the direct echellon.

5. A direct echellon (B) is formed for the purpose of advancing or refusing a flank; and the distances (*a a*) between its divisions will be guided by circumstances: A line parallel to that from which it moved, can only be resumed by the successive formation of its divisions upon one of its halted parts; and an oblique line cannot be accurately formed by the wheel of divisions from a direct echellon otherwise than by taking up distances of the preceding divisions (*c c c*) from the leading flank of the echellon (*o*) and a point (*b*) placed beyond it, as laid down in S. 122.

Of the wheeled echellon.

6. The wheeled echellon (C) possesses the double property of gaining ground to a flank in an oblique direction (*a e*), and at the same time of affording a ready and immediate formation of an accurate line (*e c*), in an advanced position, parallel to that which it quitted, by the wheel backwards of divisions on its echellon flanks.

Distinctions between open column and echellon.

7. All the divisions of an open column, A, march upon one and the same perpendicular, and are therefore easily conducted. All the divisions of a wheeled echellon,

Fig. 13. echellon, C, move on different perpendiculars, each on its own, but all of them parallel to the directing one, (*d*) and removed from each other a space equal to what the divisions cut within each other. In open column the perpendicular distance from division to division is equal to the front of the following one. In echellon the smaller the wheel is, the smaller is the perpendicular distance from division to division; but in all situations of the wheeled echellon, the oblique distance from flank to flank, (*aa*) is equal to the front of the preceding division. In open column the proper pivot flank is the directing one, and the wheels are made on it into column backward and into line forward. In echellon the inner flank (or that which first joins its preceding division when the line is to be formed forward) is the directing one, and in oblique echellon the wheels are made on it, into echellon forward, and into line backward. In open column each division preserves a distance from flank to flank equal to its own front. In echellon each preserves a distance from flank to flank equal to the front of its preceding division; and it may at any time be converted into the open column, by wheeling up its divisions till they stand perpendicular to the line which passes through all its directing flanks. An open column may in the same manner be converted into the echellon column, by wheeling back its divisions, each a named number of paces, and on either flank, according to circumstances.

Method of formation. 8. The wheel from line into open column is easily ascertained by the perpendicular halt of each division on that line: but the parallelism of the wheels into echellon is more difficult to be determined; for being

being confined to no certain portion of the circle, it cannot well be announced or executed as a direction, and therefore *a given number of paces to be wheeled by bodies of equal strength,* and which serve as so many parallel bases of formation, may be the best general order that can be given.

9. Should the companies of a battalion or more considerable body be all of equal strength, and should the outward man of each take the same number of paces on the circumference of the circle, they would, after the wheel, stand parallel among themselves; but, if those companies be unequal, they would then not be parallel to each other, and consequently not in a proper relative situation. Though such equality may be found in a single battalion, it will seldom exist in a line of battalions, and a different calculation and direction for each battalion, in proportion to its strength, would be necessary in every general change of position. This difficulty may be obviated by adopting a *practical rule* for the battalion and line, on all occasions of wheeling by companies into echellon, in order to change position, and of whatever strength

Pl. 1. the companies may be; *viz., That each covering ser-*

Fig. F. *jeant, as the case requires, having previously placed himself before or behind a given file (the 8th) from the standing flank, shall take the named number of wheeling paces from the outward shoulder of that file on the arc of the circle, and thereby become a direction for the company to wheel up to, and halt,* as in S. 114.—As eight paces from the eighth file complete the quarter circle, so four paces give the *one-eighth,* and two paces the *one-sixteenth* of the circle, all which are wheels often made from open column, or from line, to change to a position

position perpendicular, or oblique, to the one quitted; and these degrees, with the aids given by advancing or keeping back a shoulder as may be necessary, during the movement, will effect a formation in any new direction with precision.

General attentions of directing files. 10. The flank directing files of echellons, whether they are formed by the perpendicular march of divisions successively from line to the front, or by the wheel of divisions from line to the flank, should form, and preserve a diagonal line with respect to the front

Fig. 13. of divisions:—In the first case, B, the distance from flank to flank, (*aa*) depends on the interval at which the divisions are ordered to march off:—In the second case, C, such distance (*aa*) is always the same and equal to the front of the division which has wheeled forward, and which by wheeling back, would exactly fill it up. Whenever, therefore, the directing flanks of an echellon are all in the same line, and each distant from its preceding one a space equal to the front of the preceding division, such echellon is in a situation, by wheeling back, to form in line to the flank, or to take a position to the front, as in (S. 120.)

Passing obstacles. 11. In the echellon march, such division or divisions, as may meet with obstacles, will file round them without deranging the adjoining divisions, which preserve the necessary vacant spaces and distances till the broken divisions can again take their places.

Changes to the rear in echellon. 12. When a change of position or march to the rear is to be made in echellon, the battalion or line will in general *face* about, wheel into echellon, and then proceed:—or it may be ordered first to *wheel* back into echellon, then face about, and proceed as above.

13. But

Directing points.

13. But the march in echellon (more particularly that to a flank) can never be performed well unless the officer of the leading echellon shall have correct points to move upon: In the direct march the relative distance of divisions is easily preserved—but in the diagonal movement the utmost care and exactness are necessary.

14. In advancing to the right, the officers glance the eyes over the right shoulder without affecting the squareness of their front, and cover in the oblique line of officers. The reverse is the case in advancing to the left.

Superintendence of commanding officers.

15. The superintendence of the commanding officer, who should keep close on the flanks of the leading echellon, will be constantly necessary to preserve the equality of march; and to rectify any deviation he may observe in the oblique line and covering.

Change of position.

16. Each change of position, of the battalion or line, may be considered as a general wheel of the whole made on a *point*, either *in*, *before*, or *behind*, the old line. The battalion or line therefore breaks to whichever hand, and to whichever division, it manœuvres to, or is led by:—When to a flank, generally to that which is nearest to, and is first to enter any part of the new position:—When a central division determines its movement, it breaks to right or left inwards, and faces such division, which makes its change of situation on its own ground. When this *point* is *in* the old line, it must necessarily be within the battalion when single, or within a certain named battalion of a line:—Such battalion therefore will have to perform the change on a *fixed*

fixed point within itself; *viz.*, on such division, flank or central, as is already rested on that point, by making its other divisions, either by *flank* or *diagonal* marching, enter into the line: But all the other battalions will have the double operation of moving up to the new line, and then forming upon it. When this *point* is *before* or *behind* the old line, every battalion, whether single or connected, will have this double operation to perform.

Advantages of echellon in changes of position.

17. The echellon changes of position are the safest that can be employed in the presence of and near to an enemy; they are almost equal in security to the march of the line in front, or to an uniform wheel in the line, which is not to be attempted; they can be used in the most critical situations, where the filings and movements of the open column could not be risked; they are more particularly employed when the enemy's flank is to be taken by throwing the body forward, or when one's own is to be covered by throwing it backward. They have the advantage of preserving a general front during the march, and of affording a sufficient freedom of movement which in such situation is indispensable; they effect a change of position on any division of the line, either on a fixed or moving point; and at any instant the movement can be stopped, the line formed, and a sudden attack repulsed.

Fig. 72, 73. — 74.

Degrees of march.

18. The march in echellon will be, generally, in quick time, unless the commander may see occasion according to circumstances, to move in slow time. But as changes of position in echellon are always supposed to be near an enemy, where the utmost regularity and constant readiness to form line must be preserved—

preserved—it would not be prudent to attempt greater celerity than quick time.

Double march. 19. Rapid formations to line, however, from column or square upon any fixed part of a battalion, when made by the echellon movements of subdivisions or sections, may be performed at the double march: or, in echellon changes of position, when two or three divisions remain to be formed on the flank, they may complete the formation at the double march.

20. The divisions of a battalion might move from line upon the moveable pivot, without previously wheeling up the number of paces required, provided dependance could be placed upon the accuracy of officers, in giving the word *forward* at the very moment when the leading flanks are in a perpendicular direction to the point of appui in the new line. But as this cannot be relied upon in all cases, it will save time and confusion, to wheel on all occasions of echellon formation, half the angle wheeled by the division to be formed upon, (Vide S. 117.) Upon this principle of mathematical precision, therefore, all echellon formations to line from column, by sub-divisions or sections, should, strictly speaking, be made by the previous wheel backwards of divisions into echellon, on their reverse flanks, (as laid down for companies in S. 120,) in order that the leading flanks may stand precisely perpendicular to the direction in which they are to move; but, as expedition is the first object where it is easily attainable without risking confusion, the bringing forward of the shoulders of the several *small* divisions may be adopted on such special occasions, and will be found to answer every practical purpose.

79.

S. 79. Squares.

1. Squares are formed either from line (Vide S. 89 and 90.), or from column at full, half, or quarter distance (Vide S. 103 and 107.)

Hollow square, four deep. Fig. 47. 2. The hollow square, four deep, is sufficiently solid to oppose an attack of cavalry: It possesses, at the same time, the advantage of rendering the fire of all the men available to the resistance of the enemy; and of being flexible and ready to move, as well as applicable to an immediate change of formation.

Solid squares. Fig. 48. 3. When a strong battalion is formed into a solid square, a proportion of the men must be supernumerary on the flank faces, where they cannot give their fire. This square, therefore, should seldom be adopted where a battalion exceeds twenty files in each company: But as every position in which a battalion may be placed, should be susceptible of a ready formation into square, in order to avert the possibility of being surprised, or hurried by an attack of cavalry, the solid square is regulated in a subsequent section; and the close column will be found applicable to immediate resistance, by the facility with which it is formed into a solid square. (Vide S. 108.)

Squares two deep. Fig. 29. A. 4. When cavalry is not to be resisted; and when the object is only to protect baggage or treasure requiring a greater internal space, it will be sufficient to form the square two deep to the rear, (Vide S. 90.) and the same may be formed to the front as occasions may require.

Echellon of squares. 5. When battalion squares are to be formed either from line or column, composed of several battalions; the line, or the column in mass, or the line of contiguous columns, must be previously thrown into echellon of battalions; or so placed (Vide S. 142, Fig. 75. No. 26,) by the wheel of the divisions on which the square is formed, that the adjoining angles of each square shall be clear, in order to create a mutual defence from the faces, which will then flank each other at right angles. When battalions are in echellon, the degree of distance between each to enable them to form squares to flank each other, will depend upon the nature of the ground—upon the circumstances under which the formation is made—and upon the discretion of commanding officers: and, indeed, it may be immaterial for objects of defence, provided the flanking faces are clear, and that the distance does not exceed half musquet shot.

Small squares 6. When a battalion is connected in all its parts, it is not expedient that it should be broken into small squares; a square, composed of the whole, being a more efficient formation of defence. But circumstances may arise wherein a battalion moving in echellon, may be prevented from concentrating in square, upon the sudden approach of cavalry; or where two or more companies of a battalion may be separated from the main body: In these cases, an echellon of Fig. 63. grand division squares (Vide S. 123.), may be considered a formation, which affords a ready defence and mutual protection. When there is no room, for an echellon of squares, wing, or grand division squares Fig 43. from column, as laid down in (S. 103, No. 4, and No. 6,) will likewise afford an immediate resistance to to cavalry; and indeed any two bodies of a column, of equal strength, formed at half distance, may by the formation of threes, closing to their centre, and wheeling backward and forward by divisions from their centre, form a complete square. When troops are dispersed, the rallying square (Vide S. 67.), which is laid down in the fundamental instructions to the recruit, is a most important formation, when soldiers happen, from fatigue, or other causes, to be separated from a column of march, and threatened by the enemy's cavalry; and too much pains cannot be taken, to habituate the soldiers to its rapid execution. It should, therefore, be constantly practised in the drill of companies and battalions; and soldiers of different regiments should, when opportunities offer, be dispersed for the purpose of practising this formation with readiness as they happen to be near each other.

Fig. E.

S. 80. *Firings.*

1. In all movements, firing should commence after a formation, whether by companies from the flanks or centre—by wings—or by battalions; but this will depend much upon circumstances, and the discretion of commanding officers. After a march to the front, the firing of a battalion should generally commence from the centre, and not from the flanks: In successive formations, it may begin from the division on which the change of position is made.

Time of firing by divisions, &c.

2. In firing by companies from right to left—from left to right—from flanks to centre, and from centre to flanks, the leaders of each will step out one pace

to the front and face inwards; and will be careful to observe the following directions. To *make ready* when the previous division *fires;* to see that their companies bring up their pieces regularly to the *present*, and to preserve the pause of slow time, *viz.*, the seventy-fifth part of a minute, between each of the words "*Ready*"—"*Present*"—"*Fire.*"

By wings. 3. In firing by wings, one wing will *make ready* the instant the other is *shouldering*;—the commanding officer of the battalion will fire one wing, the second in command the other: Wings may offer a destructive cross fire by the oblique present inwards.

File firing. 4. File or independent firing from the right or left of wings, or from the right or left of sub-divisions, should be practised frequently, as being the most essential and usual mode of firing upon actual service.—It should be practised in the formations to line, after an advance from the flanks or from the centre of battalions. (Vide S. 86.) The divisions upon which the formation to line is made upon such occasions, will commence the fire, while the remaining divisions are in movement; and the latter will take it up, successively, as they are formed in line.

File firing during formations. 5. It is most essential that battalions should be well practised in formations while divisions already formed are firing,—for precise dressing will then be difficult; and officers commanding divisions must therefore give their utmost attention to the true parallel direction of their divisions upon the line; the outer section of each division will always reserve its fire until the officer of the next division shall give the word

word *Eyes front* to his men after quickly dressing them.

Attentions in file firing

6. Great care must be taken in file firing, that it is not hurried; and that the men *present* deliberately, bringing up the firelock gradually and looking at their object before they fire,—otherwise it will lose all its effect against an enemy. The value of a soldier's ammunition, and a jealousy of its expenditure without effect, must be carefully inculcated;—for in proportion as a cool and well directed fire serves to distract and throw an enemy into disorder, so is a wild, confused, and hurried fire (which is always without effect), calculated to give him confidence and a contempt for his opponent:—It is impossible, therefore, to labour too much at giving to soldiers the habit of steady, cool, and effective firing: They should be practised to aim from a hollow at objects on high ground, and from the latter at objects in a hollow, as well as in all the different inequalities of surface. They should likewise be practised, as frequently as possible, to fire with ball: but the difficulty of finding situations in which this kind of practice could be carried on, must of course retard the perfection at which the soldier should arrive: much, however, may be done by careful instruction, even with blank cartridge; and commanding officers must direct their special attention to the detailed instructions upon this important subject, which are annexed to the manual and platoon exercise.

No hurry in file firing

7. According to these principles, file firing must be conducted slowly and deliberately. Each file must bring up the piece to the present at the same time

but the rear rank man must not fire until the front rank man has fired; and the front rank man must always reserve his fire until the file he follows has fired: If this is carefully attended to, no hurry, and consequent loss of fire and intermission of firing, can ensue.

Firing in square. 8. In firing in square the two front ranks should come to the kneeling position (on preparing to receive cavalry) without cocking; because, when they are not required to fire, which can seldom be the case from each of the four faces at the same time, there is an awkwardness in half-cocking while in that constrained position. Each face will therefore fire when required, and will cock as usual at the previous word "*Ready*." The standing ranks in square will fire independently from the right of faces, as already explained in No. 4.

Street firing. 9. An effective successive fire of divisions may be given by a column at open, half, or quarter distance, when formed in a street, or in narrow ground where deployment is impracticable:—It will be performed in the following manner:—The leading division will stand fast, and the remaining divisions will *form four deep*:—The leading division will give its fire, and *shoulder arms,—form four deep,* and go to the *right about,* passing through the intervals of the rear companies. The moment the front of No. 1. is clear, it will form two deep, and give its fire,—*shoulder arms,—form four deep,*—go to the *right about,* and retire as before. Fig. 51. As each company gains the rear in succession, it *halts, fronts,* and *loads*. In this movement, when each company retires through the intervals of those in its rear, the officer commanding it places himself directly

directly in *front* of his pivot file, and passes with it through the interval, that he may not disturb the officers on the flanks of the other divisions. (Vide S. 111. No. 2.)

This mode of firing is particularly applicable to the advance of a column into a fortified or open town, but it may be adapted to a variety of occasions, where the nature of the ground or other circumstances render it desirable.

FORMATION OF THE BATTALION.

Strength of the battalion.

THE battalion is composed of eight companies, viz., 1 Grenadier, 6 Battalion, 1 Light.

Each company consists at present of 3 Officers, 3 Serjeants, 3 Corporals, 1 Drummer. 69 Privates.

Formation of the battalion.

When the companies join, and the battalion is formed, there is to be no interval between any of them, grenadier, light company, or other; but every part of the front of the battalion should be equally strong.

Each company which makes a part of the same line, and is to act in it, must be formed and arranged in the same manner.

Position of the companies in battalion.

The companies will draw up as follows from right to left according to the regimental rank of their respective

spective captains; viz.—grenadiers;—1st captain;—3d captain;—5th captain;—6th captain;—4th captain;—2d captain;—light company.

Divisions. The battalion will be told off as follows, viz., four grand divisions;—eight companies,—sixteen sub-divisions,—thirty-two sections, when sufficiently strong to be so divided, otherwise twenty-four, for the purposes of march.—The battalion is also divided into right and left wings.

The battalion companies will be numbered from the right to the left, 1. 2. 3. 4. 5. 6.—The sub-divisions will be termed right and left of each;—the sections will be numbered 1. 2. 3. 4. of each, &c.—The grenadier and light companies will be numbered separately in the same manner, and with the addition of those distinctions. These several appellations will be preserved, whether faced to front or rear.

Companies equalized. The companies must be equalized in point of numbers, at all times when the battalion is formed for field movement; and could the battalions of a line also be equalized, the greatest advantages would arise; but though from the different strengths of battalions this cannot take place, yet the first requisite always must, and is indispensable.

Formation of the battalion at close order. Ranks are at the distance of one pace, except the third or supernumerary rank, which has three paces.

All the field officers and the adjutant are mounted.

The commanding officer is the only officer advanced in front, for the general purpose of exercise when the battalion is single; but in the march in line, and in the firings, he is in the rear of the colours.

The lieutenant-colonel is behind the colours, twelve paces from the rear rank.

The

The first major is six paces in the rear of the second battalion company from the right flank: The second major at the same distance in the rear of the second battalion company from the left flank: The adjutant at the same distance in rear of the colours.

One officer is on the right of the front rank of each company, and one on the left of the battalion; all these are covered in the rear rank by their respective serjeants; and the remaining officers and serjeants are in a third rank behind their companies.

The colours are placed between the third and fourth battalion companies, both in the front rank, and each covered by a non-commissioned officer, or steady man in the rear rank. One serjeant is in the front rank betwixt the colours; he is covered by a second serjeant in the rear rank, and by a third in the supernumerary rank. The sole business of these three serjeants is, when the battalion moves in line, to advance and direct the march as hereafter mentioned. The place of the first of those serjeants, when they do move out, is preserved by a named officer or serjeant, who moves up from the supernumerary rank for that purpose.

Use of the third or supernumerary rank. The third rank is at three paces' distance when halted or marching in line. When marching in column, it must close up to the distance of the other ranks. The essential use of the third rank is, to keep the others closed up to the front during the attack, and to prevent any break beginning in the rear: on this important service, too many officers and non-commissioned officers cannot be employed.

The pioneers are assembled behind the centre, formed two deep, and nine paces from the third rank.

NOTE.—*When the battalion consists of ten companies, the colours will be placed between the fourth and fifth battalion companies.*

The

The drummers of the six battalion companies are assembled in two divisions, six paces behind the third rank of their first and sixth companies.—The grenadier and light company drummers and fifers are six paces behind their respective companies.

The musicians are three paces behind the pioneers in a single rank, and at all times, as well as the drummers and pioneers, are formed at loose files, only occupying no more space than is necessary.

The staff of surgeon, assistant-surgeon, and quartermaster, are three paces behind the music.

Officers. In general, officers remain posted with their proper companies; but commanding officers will occasionally make such changes as they may find necessary.

Replacing serjeants. Whenever the officers move out of the front rank, in parade, marching in column, wheeling into line, or otherwise, their places are taken by their serjeant coverers, and preserved until the officers again resume them.

When the line is halted, and especially during the firings when engaged, the serjeant coverers fall back into the third rank, and observe their companies.

When the Battalion takes Open Order.

Rear Ranks take Open Order. At the word *Order*, officers recover swords, and two aids are placed with their flags erect, on the right, and one on the left of the battalion, three paces in front: They are corrected in the proper line of covering by the first major. The flank men on the right of the rear rank of each company step briskly back one pace, to mark the ground upon which the rank is to halt; they face to the right, and cover as pivots, corrected by the serjeant-major on the right.

At

March. At this word, the rear rank falls back on pace, dressing by the right: The leaders of companies march three paces obliquely to the left, so as to place themselves in front of, and opposite to, the second file; then instantly face to the right, and cover the points afforded by the adjutants' aids: The officers with the colours march forward three paces, and cover to their right: The other officers pass through the intervals, and cover to the right; the lieutenants taking post on the second file from the left; the remaining offiers (if there are more than two subalterns) covering the centre of the second and third sections: If there are only two subalterns, the second subaltern will cover opposite the centre of the company.

The music pass through the centre of the battalion, and form in rank entire between the colours and the front rank. The pioneers fall back six paces behind the centre of the rear rank: The drummers take the same distance behind their divisions. The first major places himself on the right of the line of officers: The second major on their left: The adjutant on the left of the front rank: The staff, viz., the surgeon, assistant surgeon, and quarter-master, place themselves on the right of the front rank of the grenadiers, at one pace distance. The colonel and lieutenant-colonel (dismounted) advance four and two paces before the colours. The serjeant coverers move up to the front rank, to preserve the intervals left by the officers.

The

The whole remain in this position until the first major, who has corrected the covering of the officers, orders the aids to *lower their flags;* and upon this signal, the officers face to the *front*, and drop their swords across their bodies: The aids retire to their places.

The whole thus arrive at their several posts, and the battalion remains formed in this parade order, to receive a superior officer. When the battalion is reviewed singly, the division of drummers may be moved up and formed two deep on each flank of the line: The pioneers may be formed two deep on the right of the drummers of the right, and the staff may form on the right of the whole.

When the Battalion resumes Close Order.

Rear Rank take Close Order.

The lieutenant-colonel, officers, colours, staff, music, face to the right.

The drummers and pioneers (if on the flanks) face to the centre.

The serjeants (if in the front rank) face to the right.

March.

The rear rank closes within one pace.

The music marches through the centre interval.

The serjeants, drummers, pioneers, &c. &c., resume their places, each as in the original formation of the battalion in close order.

The officers move through and into their respective intervals, and each individual arrives, and places himself properly at his post in close order.

Colours. When the battalion wheels by companies or subdivisions

divisions to either flank into column; both colours and the file or directing serjeants always wheel to the proper front, and place themselves behind the third file from the new pivot.

Colour reserve.

There is no separate colour reserve; the pioneers, music, &c., sufficiently strengthen the centre: but in the firings the two files on each side of the colours may be ordered to reserve their fire.

EVOLUTIONS OF THE BATTALION.

The rules laid down and explained in Part II. for the formation of columns from line, and line from columns,—for marching in alignement,—wheeling upon fixed and moveable pivots,—diminishing and increasing the front of columns,—and countermarching the divisions of a column, apply equally, in all the detail of instruction, to the company in battalion: Commanding officers will, therefore, conduct their battalion drill, upon all those points, with reference to the said rules, and to the instructions cootained in the general principles; as it is not intended to frame any specific sections in the following manœuvres, to explain those formations as applied to battalion movements.

MOVEMENTS OF THE BATTALION FROM LINE.

(Plates III, IV.)

S. 81. *When the Battalion halted, and correctly dressed, is to advance in Line.*

THE BATTALION WILL ADVANCE. — 1. On the CAUTION, the centre serjeants will advance and take their direction, corrected by a mounted, or the commanding officer, as pointed out in the General Principles, S. 70.

QUICK MARCH — The line of direction being thus ascertained, at the word QUICK MARCH, the whole battalion instantly step off, the eyes directed full to the front, the files of each wing preserving a light touch inwards; and shoulders, as well as heads, kept square to the front.

HALT. — When the line halts the directing serjeants will resume their place in the battalion, and be in readiness to move out again, if required to advance after firing, or dressing.

2. *When the battalion is to retire*, it must be previously dressed, as instructed in S. 74. No. 11, with the same precision as when it was o advance; and the direction of the march must be ascertained with the same accuracy. At the word THE BATTALION WILL RETIRE, the directing serjeants face about, and the same arrangements are made, as already directed for the advance in S. 70.

Fig. 15. — 3. While advancing in line, (*a b*) the battalion may form to either flank by the divisions wheeling to the right (or left) on the moveable pivot, and forming

ing on the flank company (*b c*) (which will be halted in the direction of the new front) by the echellon march of divisions.

S. 82. *When a Battalion advancing in Line is to Charge.*

PREPARE TO CHARGE. Upon the caution PREPARE TO CHARGE being given, the front rank will bring their arms to the long trail, and the rear rank to the port, without permitting the motion to alter the square position of the body, or the regularity of the step.

CHARGE. And upon the word CHARGE, which ought soon to follow the caution, the front rank only will come to the charge, the whole battalion stepping off at the same moment at the double march.

HALT. When the battalion HALTS, both ranks will come to the shoulder, and the whole will dress by the centre and commence independent firing, or advance or retire from the right of companies according to circumstances, or the discretion of the commander.

S. 83. *When the Battalion moving in Line passes a Wood, or other Impediment, to Front or Rear, by the Flank March of Companies by Threes.*

1. *If to pass a wood, or other embarrassed ground to the front.*—When it is found necessary to break the battalion, the commander wil

PASS BY THREES TO THE FRONT. *Threes right, Left wheel.*

Fig. 14—A.

will order it to PASS from the right of companies to the front, on which each company officer orders his company *Threes right, left wheel,* and passes on as fast as the difficulty of the ground will allow him, endeavouring to preserve a relative distance from the left as being the head of the column, or from the other flank if particularly so ordered. Each officer on arriving at the farther edge of the wood will *halt* his company, and remain till the others are come up, and till the whole are ordered to march out and form in battalion; which will generally be done by standing in open column the left in front, dressing pivot flanks, and wheeling up into line.—Or, if the companies form separately on the edge of the wood, they will march out and join in the battalion.

Fig. 14—B. PASS BY THREES TO THE REAR. *Threes left, Right wheel.*

2. *If to pass to the rear.*—When the battalion retiring in line (B), arrives at the point where it must break, it is ordered to PASS COMPANIES BY THREES.—The leader of each gives his word *Threes left, right wheel,* and proceeds as above directed; the heads of threes are regulated from the left; and after quitting the wood, at an ordered distance they HALT, FRONT into column the right in front, and WHEEL to the left into line.—The line then again retreats if necessary.

HALT, FRONT.

Fig. 16.

3. *If a battalion in first line passes through a second which advances and relieves it*—the second (*b*) marches up to within 12 paces of the first (*a*) and halts.—The battalion of the first

PASS BY THREES TO THE REAR. *Threes Right. Right Wheel. Quick March.*

first then receives the word from the commanding officer, PASS BY THREES TO THE REAR. Each leader gives his word *Threes right, Right wheel, Quick March,* and proceeds at a quick pace to the rear through the second line, which, whenever the head of a division presents itself, throws back as many files as are necessary to give it passage, and again immediately moves up; the retiring divisions, who are regulated by their left, at any ordered distance HALT, FRONT into column the right in front, and wheel up to the left into line (a).

HALT, FRONT.

PASS BY THREES TO THE REAR. *Threes left, Right Wheel.*

4. *When the second line does not advance to relieve the first,* the battalion of the first line retires, and when it comes within 12 paces of the second, it then receives the word to PASS BY THREES TO THE REAR; each leader orders *Threes left, Right wheel,* and proceeds as before directed; the column when halted and fronted, having its right in front.

Circumstances may require, that the companies should PASS from their proper left instead of the right, in which case the leaders will shift and conduct such left, until the line is formed, when they will again resume their proper places.

Fig. 17.

PASS BY THREES TO THE FRONT.

5. *If a battalion in second line passes by threes to the front through a first line,* it will advance within 12 paces (*a*) of the first line (*b*). On the command to PASS BY THREES TO THE FRONT; each company leader will give his word *Threes Right, Left wheel,* and move on through the first line, which makes openings for it. When the rear of the files has

 passed

passed, the battalion will be ordered HALT FRONT in column the left in front—WHEEL into line (a)—and may then advance.

Fig. 18.

6. *If a battalion in second line advances and passes in front, through a first line which it is to relieve.*—The first line (*b*) will, at the necessary instant, FORM FOUR DEEP, and the advancing battalion will also form four deep on the march (*a*), (Vide S. 47. No. 5) and pass through the intervals to the right, and when passed will RE-FORM TWO DEEP: (a). In this manner lines may advance or retreat through each other, as circumstances may require, each file passing to the right of the one opposed to it.

FORM FOUR DEEP.

RE-FORM TWO DEEP.

S. 84. *When the Battalion retires by alternate Companies.*

RIGHT COMPANIES, HALT, FRONT.
LEFT COMPANIES, HALT, FRONT.

RIGHT COMPANIES. ABOUT FACE, QUICKMARCH. HALT, FRONT.

LEFT COMPANIES. ABOUT FACE, QUICKMARCH. HALT, FRONT.

The right companies stand fast, or HALT, FRONT if the battalion is already in motion.—The left retire in line a given number of paces and HALT, FRONT: on which the right companies retire in the same manner beyond the left, and HALT, FRONT: In this way they proceed till the battalion is ordered to form. One colour remains on the flank of its proper company in each line, and directs its movement,

movement, for which purpose a serjeant will advance 6 paces before it during the march. The eyes of each line remain directed full to the front, and the touch to the colour; and officers are on the inward flanks of their companies.— Each line has a command. The light infantry may be divided in the intervals of the first line, retire with it, and change to the other line whenever it becomes the advanced one: in this situation they cover the retreat, and may fire.

S. 85. *When the Battalion advances or retires by half Battalions, and fires.*

LEFT WING, HALT. QUICK MARCH.

RIGHT WING, HALT, READY. PRESENT. FIRE. QUICK MARCH.

LEFT WING——HALT. READY, &c.

1. *If the battalion is in march, and advancing.*—The left wing HALTS when ordered, and the right one continues to move on 15 paces at which instant the word MARCH being given to the left wing, the right at the same time is ordered to HALT, to fire and load, and the left marches past them, till the right wing being loaded and shouldered, receives the word MARCH, the other wing HALTS, fires, &c., and thus they alternately proceed.

 2. *If*

RIGHT WING, HALT, FRONT. LEFT WING, HALT, FRONT. RIGHT WING, { READY. PRESENT. FIRE. ABOUT FACE. QUICK MARCH. LEFT WING, HALT, FRONT. LEFT WING, { READY. PRESENT. FIRE, &c.	2. *If the battalion is in march, and retiring.*—The right wing is ordered to HALT, FRONT, and when the left one has gained 15 paces, and receives the word HALT, FRONT, the right wing is instantly ordered to FIRE, to LOAD, to FACE about, and march 15 paces beyond the left, where it receives the word HALT, FRONT, on which the left wing gets that of FIRE, and in the same manner alternately proceeds, every due despatch being made in reloading.

There must be a commander for each half battalion.

One colour, before which, a directing serjeant advances six paces, remains on the inward flank of each half battalion, towards which the men continue to touch; the officers of the right wing to be on the inward flank of their companies.

The make ready, present, fire, of the advanced wing, is instantly to succeed the march of the other advanced wing, or the halt front of the retiring wing.

S. 86. *A Battalion formed in Line may have to move to attack, or to pass a Bridge, or short Defilé, to the front, from either Flank, or from the centre.*

RIGHT (OR LEFT) DIVISION TO THE FRONT.	1. *If from a flank, by companies, or sub-divisions.*—The caution will specify from which flank the advance is to be made;

REMAINING DIVISIONS RIGHT (OR LEFT) SHOULDERS FORWARD. QUICK MARCH.

Fig. 19—A.

Right (or left) shoulders forward.

Forward.

made; and upon the word QUICK MARCH, (after the annexed command) the flank company, or sub-division, moves on in the line proposed (*dd*). The others advance in corresponding divisions at the same time to the right or left, according to the flank in advance; the commander giving the word, FORWARD, when the divisions shall have wheeled square into column, at which time the leaders will change to the left flanks, if the movement is from the right; and while thus wheeling into column, the division (*a*) will be gaining sufficient ground to the front to enable the others to follow by command of their leaders *Right (or left) shoulders forward,* in regular succession, without loss of time or ground.

If the movement be made to pass a bridge or defilé, the front of the column will correspond with the breadth of the bridge or defilé.

FORM LINE ON THE FIRST DIVISION. REMAINING DIVISIONS RIGHT (OR LEFT) SHOULDERS FORWARD. Q. *or* D. MARCH.

Fig. 20—B.

Right (or left) Shoulders Forward.

Forward.

Halt, Dress up

2. *When the column arrives at the point where the line is to be re-formed*—it is halted (B). The caution is given, and (supposing it a column of sub-divisions or sections) the formation to line may be rapidly made upon the front division (a), by each of the others wheeling at once to the flank, on the moveable pivot; and proceeding under the direction of their respective officers, *Right or left shoulders forward,* to form line (*e* a), upon the covering serjeants of companies, who take points of distance (*s s*) (Vide S. 74. No. 7.), the leaders of companies giving the word *Halt, Dress up,* from the point of appui.

When

When the line is formed from column of companies, the echellon formation should be conducted as laid down in S. 120, (Vide S. 78, No. 20,) or the column may close (a *c*) and deploy.

Fig. 79.

Fig. 19—B.

TWO CENTRE SUB-DIVISIONS TO THE FRONT. REMAINING SUB-DIVISIONS RIGHT AND LEFT SHOULDERS FORWARD. QUICK MARCH.

Fig. 20—A.

FORWARD.

3. *If the advance is from the centre.*—Upon the caution being given, the colours, &c., will fall back to the rear of the centre; and upon the word QUICK MARCH (after the annexed command) the whole will be thrown at once into movement, the commander giving the word, FORWARD, when the divisions have wheeled square into column; at which time the leaders of the left wing divisions will shift their flanks, and each sub-division will have an officer at the outer flank *(ss)*. The sub-divisions of the right and left wings will thus form into column respectively, on the centre sub-divisions (*a* a); each wing being conducted in the same manner, and upon the same principles, as described in the formation to the flank. During the advance, the interval between the centre sub-divisions will be closed to the directing flank, leaving a serjeant in the centre between each sub-division.

A serjeant will always be placed between each of the divisions of a double column, to mark the interval between each. The left is always to be considered the governing flank by which the column is to march, unless ordered to the contrary.

Double columns, unless ordered to the contrary, will be formed at the distance of the divisions of which each column is composed: And they may be formed from the halt in line, upon any two centre divisions, like any other columns, as in S. 93, or S. 104, No. 1.

FORM LINE ON TWO CENTRE SUB-DIVISIONS REMAINING SUB-DIVISIONS RIGHT AND LEFT SHOULDERS FORWARD. Q. *or* D. MARCH.

Fig. 20—B.

[FORWARD.

4. *When the double column arrives at the point where the line is to be formed*—it is halted, the caution is given, that the column will FORM LINE ON THE TWO CENTRE SUB-DIVISIONS (aa), and upon the word QUICK MARCH, the latter open out by the side step, and the colours will resume their place in the interval. The rear sub-divisions at the same time move off, and are brought into line by their respective leaders conducting them, *Right and left shoulders forward*, from the inward flanks, to which they shift when the commander gives the word FORWARD; the sub-divisions of each wing forming line (*ee*) upon the centre, and the covering serjeants of companies, as already described, upon a flank.

Fig. 20—C.

In these formations the column may be closed to quarter distance (*co*), and the line formed by deployment, according to the nature of the ground, and the discretion of the commander; or the line may be formed upon the flanks or centre by deployment without closing, if the ground is clear. —The rear subdivisions facing outwards by threes, and moving to their respective parallel positions with the intended line, they will then *Front turn*, and form upon the coverers. (Vide *S.* 74, No. 7.)

RIGHT WING LEFT SHOULDERS FORWARD.

5. *If, after having gained the desired point to the front, it be required to form line to the right from the double column.*—The right subdivisions are formed up into

Halt, Dress.
LEFT WING.
Q. *or* D. MARCH.
Fig. 23.
Halt, Dress.
Left shoulder forward.
Halt, Dress.

into line by the word LEFT SHOULDERS FORWARD, and are halted and dressed in ine (c *b*) by their respective officers; the left wing sub-divisions (*dc*) at the same time advance and form line (cd) on the right wing, being conducted by their respective leaders, who change their flank, and bring the *left shoulder forward* into line, dressing upon the covering serjeants and battalion points.

In the same manner line may be formed to the left, or the double column may be previously halted, the divisions of one wing to WHEEL INTO LINE, the other wing to advance, and form upon it in successive divisions.

FORM LINE ON THE TWO CENTRE SUB-DIVISIONS FACING TO THE REAR.
Fig. 21.
Right and left face, right countermarch, Quick march.

6. *If the formation to line from double column be required to the rear.*—Upon the caution, the covering serjeants of the two centre companies will change places by the rear of their divisions; and they will take sufficient ground to each flank, to leave room for the colours (which will countermarch) to resume their position in line: The leaders of the two centre sub-divisions will face them outwards, and will countermarch them by files, changing places, and forming on their respective covering serjeants.

BY SUB-DIVISIONS THREES OUTWARDS. RIGHT COUNTERMARCH.
Q. *or* D. MARCH.
Rear turn, Halt, front, dress up.

The remaining sub-divisions will face outwards by threes, and will proceed, after the countermarch, until they are clear for deployment on the rear, when they will *Rear turn*, and *Halt, Front, Dress up*, upon the covering serjeants of companies. Vide S. 74, No. 7.

7. *If*

Fig. 22.

FORM LINE ON THE REAR BASE FACING TO THE REAR. WINGS WILL CHANGE PLACES. SUB-DIVISIONS THREES OUTWARDS. RIGHT COUNTER-MARCH AND DEPLOY. Q. *or* D. MARCH.

Halt, Front, Double March. Halt, Dress up.

7. *If the formation to line be required upon the alignement of the two rear subdivisions, facing to the rear.*—The caution will be given, and the two serjeants of the centre companies will change places, and, as base points, will advance along the flanks, and take up the alignement of the rear divisions (*cc*), and the two centre subdivisions will countermarch by files, and change place by command of their respective leaders, as before. The remaining sub-divisions will then face outwards by threes, will countermarch to the right, and will successively deploy (Vide S. 113.) upon the two centre sub-divisions, which will advance at the *Double March* through the centre, by command of the officer on the left flank, the moment the opening of the other sub-divisions (*ss*), will admit their front; and they will halt at their base points (*c c*), the other sub-divisions receiving the word *Halt, Front*, and forming upon the centre as in any other deployment to a front alignement.

Upon the caution to advance in all these occasions, the light infantry will form in two divisions, and will extend and cover the flanks, or front, to support the attack, or to cover the passage of the bridge or defilé.

S. 87. *A Battalion formed in Line may have to retire over a Bridge or short Defilé, or to retreat from a flank or from the centre.*

1. *If from a flank*—the battalion will change its front to the rear upon the centre (vide S. 91.) and will then advance according to the width of the defilé, as instructed in last section. If there should not be time for a previous change of front, and if it be intended to resume the former front after passing the defilé, the battalion will be put to the right about, and will retire with the rear rank in front, in the same manner as already explained.

BY SUB-DIVISIONS RIGHT AND LEFT BACKWARDS WHEEL, QUICK MARCH.

Fig. 24—A.

RETIRE FROM THE CENTRE. QUICK MARCH.

Right and left shoulders forward.

2. *If the retreat is from the centre.*—Upon the caution, the colours and centre serjeants will advance a few paces, and countermarch; and the battalion will break into column backwards by sub-divisions, (the colours forming in rear of the right centre subdivision,) the right wing on the right, and the left wing on the left, (*dd*), thus forming two columns facing the centre, and upon the word RETIRE FROM THE CENTRE, the officers will change to the reverse flanks; upon the word QUICK MARCH, the whole step off together, the two centre sub-divisions immediately wheeling the quarter circle (*c c*) by word from their respective leaders, *right and left shoulders forward*, and moving on to the rear (B): the other sub-divisions following in double column (*a a*) in succession on the moveable pivot.

HALT.
THE WINGS WILL CHANGE PLACES.
INWARDS FACE QUICK MARCH.

Halt, Front,

Left, Dress.

3. *If it be intended to form line to the former rear,* the double column thus formed will be halted, to rectify its inversion; for which purpose the wings will change places by facing inwards, and passing each other; and the respective leaders of sub-divisions will follow in the rear, until they reach the centre serjeant, when they will give the word *Halt, Front,* and then proceed to the outward flanks; the officer on the left giving the word *Left, dress,* that the sub-divisions may dress to that pivot: or it may be done on the march by the word INWARDS TURN, the leaders of sub-divisions, as before, giving the words *Front, turn,* when they reach the centre serjeants, who mark time at the word INWARDS TURN.

The double column will then be in its proper formation, and ready for any ulterior movement to its new front:

4. *But should it be required to re-form the line to the former front* after passing the defilé, the wings will not change places, but the column will be halted, and the sub-divisions will countermarch on their own ground; the line will then be formed on the centre, by deployment to the rear. Or

Fig. 24. B. upon the caution, the two centre sub-divisions will countermarch upon their own ground, the colours taking post between them: The other sub-divisions will be FACED TO THE RIGHT ABOUT, and will be wheeled RIGHT AND LEFT SHOULDERS FORWARD, then FORWARD when in a true echellon direction with the alignement of the centre, upon which they will form. (Vide S. 118.)

Upon the original caution to retire, either from flanks or centre,

centre, the light infantry will extend in two divisions from the flanks, to cover the retreat.

S. 88. *When a Battalion formed in line is successively to march off in Column of Divisions to a Flank.*

Fig. 25.
THE BATTALION WILL MOVE IN COLUMN OF DIVISIONS (OR SECTIONS) FROM THE RIGHT FLANK.
Threes left, left wheel, Front turn.

1. *If the movement is along the rear and from the right flank.*—Upon the caution being given the leading division will *face* by *threes left, left wheel, Quick March,* and will march out perpendicular to the line, the extent of its own front (*b*); it will then receive from its leader the words *Front turn.*—The division next it will face in the same manner, and will disengage its head towards the column; and when the leading division (b) passes, it will receive from its leader the words *Quick March, Front Turn;* and thus division after division (*dd*) will follow in column at their true distances.

Threes right, Right wheel.

2. When the movement is from the left flank of the battalion, the same operation takes place, except that the divisions face *threes right, right wheel,* and follow in succession as before.

It is to be observed, that marches made in this manner can seldom be required along the FRONT; when they are, it is the same operation, only disengaging and marching out by divisions to the front.

S. 89. *When the Battalion halted in line, shall be required to form a square or oblong four deep on the centre.*

Fig. 27—A.
ON THE TWO CENTRE SUB-DIVISIONS FORM SQUARE. REMAINING COMPANIES RIGHT ABOUT FACE.

Q. or D. MARCH. *Right and left shoulders forward.*

1. On the word *Form square,* the colours and centre serjeants fall to the rear, the two outer sub-divisions (*aa*) face inwards, and the leading files disengage to the right and left; a serjeant from each of the two flank companies will post himself at the proper distance, square with the flanks of the two centre sub-divisions, and mark the rear angles (*ss*): The remaining companies of the battalion face to the RIGHT ABOUT, and the two flank companies at the same time face by threes inwards, and upon the word QUICK or DOUBLE MARCH, the two centre sub-divisions close upon the interval left by the colours, while the two sub-divisions which were faced inwards, file in rear (aa) of those in their front, and the remaining companies move at once by sections, inclining forward their outward shoulders, those of the right wing, to the left, and those of the left wing, to the right, by word from their respective leaders, each of whom is to conduct his section to the relative situation it is to fill in the square; *viz.*, the left sections of No. 2, are *halted* and *fronted* by its leading officer, when three paces within the centre sub-divisions, the proper left flank of the left section being perpendicular to their rear; The right sections of No. 2, follow the left and cover in the rear, and are *halted* and *fronted* perpendicular to the flank of

Halt, Front.

of the centre, the left sections immediately closing upon them. In the same manner, No. 1 completes the right face of the square; while Nos. 5 and 6, by a similar march form its left face, the left sections covering in rear of the right; and the sub-divisions of the two flank companies are conducted by their respective leaders to form the rear face, the right sub-divisions covering the left of the grenadiers, and the left sub-divisions covering the right of the light company, each receiving the word *Halt, Rear form* when in line of the rear face, and they will dress by the left on the serjeants who are to occupy the rear angles of the flank faces.

The square will thus be in readiness to resist cavalry, as explained in the formation from column, vide S. 107. When this formation of square upon the centre has been well practised in quick time, it may always be formed at the double march; the rapidity of the formation, consistently with order, being most essential upon the approach of cavalry.

In this manner any wing of a battalion, consisting of eight sub-divisions, may form a square upon the two centre sections.

RE-FORM LINE.

Fig. 27—B.

Q. *or* D. MARCH.

Halt, Front, Dress.

2. *Reduce the square and form line.*—On the caution, the flank companies face outwards by threes, the rear sub-divisions (*b b*) of the centre, face outwards in file, and on the word QUICK OR DOUBLE MARCH, they file rapidly into line, (bb) the words *Halt, Front, Dress*, being given by their officers, while the two centre sub-divisions open to the right and left, to leave room for the colours and centre serjeants; the flank

flank and other companies step off at the same time, the former move by threes to their respective places; and the latter are brought into line by the echellon march (*c c c c*) of sub-divisions, the moment the two sub-divisions of the centre shall form. The companies are *halted* and *dressed* upon the centre, and the covering serjeants, by their respective leaders.

FORM DOUBLE COLUMN OF SUB-DIVISIONS

Fig. 27—C.

3. *But should it have been previously necessary to move the square*, it will for such purpose be reduced to column at quarter distance; and on the caution, the rear sub-divisions of the side faces will fall back to section distance, and the pivot men of the sub-divisions of those faces will face inwards.

QUICK MARCH.
Halt.

Right about face.

On the word QUICK MARCH the sections which formed the side faces wheel backwards; and at the same moment the two sub-divisions of the front face, and four sub-divisions of the rear face (*cc*, *aa*) advance to quarter distance, are *halted*, and the latter (aa) come to the *right about*, by words from their respective leaders:—The wings will open out by the side step from the centre, to leave room for a serjeant between the sub-divisions, as in S. 86, No. 3.

The whole will now stand a double column of sub-divisions at quarter distance, and may be moved in any direction like any other column. The readiness and rapidity with which square is reduced to column and re-formed into square when required,

required, obviates the necessity of explaining a march in square. Should there have been any thing within the space of the square to protect, it can form on the reverse flank of the column during the march.

4. *When the line is retiring, the square may be formed at once without halting.* Upon the word, ON TWO CENTRE SUB-DIVISIONS, FORM SQUARE, they will halt and front; the outer subdivisions will immediately *turn inwards,* and form upon them; and the formation will proceed, as already directed, at the DOUBLE MARCH.

Fig. 28—A.
FORM SQUARE FOUR DEEP ON THE TWO CENTRE COMPANIES.
INWARDS, AND REAR FORM FOUR DEEP.
Q. *or* D. MARCH.

5. *A complete square, four deep, may be rapidly formed from line in the following manner.*—Upon the caution, the colours and centre serjeants will fall back, and upon the word INWARDS, AND REAR FORM FOUR DEEP, MARCH, the two centre companies form inwards; and upon the word QUICK or D. MARCH, they will close to the centre, and the files will halt and front as they close (*cc*); while the remaining companies, which have formed four deep to the rear, will proceed into square, each company receiving from its leader the word *Right and Left Shoulders forward.* The files will close to the inward flanks of divisions during the march, and will form in square as follows:—The 1st and 2d company will form the right face, the 5th and 6th the left face, and the flank companies will form the rear face. Each company, when in its place, in square, will receive the word *Halt, Front,* from its leader.

Right and left shoulders forward.

Halt, Front.

The two centre companies must always close their fours at the

the *Quick March* though the *Double March* may be ordered for the rest of the formation.

Should it be necessary to move in square, the faces will face to the direction to which the movement may be ordered.

REFORM LINE.
Fig. 28—B.

Q. *or* D. MARCH.

Right and left shoulders forward.

Halt, Dress.

6. *When the square is to re-form line.*—Upon the caution being given the front face will open out (*aa*) from the centre and form two deep, as they open, colours and centre serjeants resuming their place; and upon the word *Quick March* the remaining companies will move off, receiving from their respective leaders the word *Right and left shoulders forward;* and during the march (*cc*) they will open out, and will be brought up two deep into line, forming upon their covering serjeants as usual.

Fig. 30.

7. *If when the light infantry is detached, it be required to form a square from line of seven companies.*—The formation will go on upon the two centre sub-divisions, as in the first paragraph of this section: No. 2 and No. 5 companies, with the left sub-divisions of No. 1 company, and the right sub-division of No. 6, company, will move in the same manner by sections, and form the right and left faces of the square; while the right sub-division of No. 1, and the left sub-division of No. 6, will form threes inwards, and will move to the rear face, forming two deep to the rear: and the grenadiers will form threes to the left, and will move to complete the rear face.

8. *When it is required to form double column.*—The right sub-division of No. 1, and the left sub-division of No. 6, and the grenadier company, will take their distances to the rear; and the formation will take place as in the third paragraph of this section, the grenadier company remaining complete at quarter distance.

In the same manner a square may be formed from a line of nine companies.

S. 90. *When the Battalion forms a Square, or Oblong, two deep, to protect Baggage or Treasure against Infantry only.*

Fig. 29—A.
FORM SQUARE TWO DEEP ON THE TWO CENTRE COMPANIES.
REMAINING COMPANIES RIGHT ABOUT FACE.
RIGHT AND LEFT SHOULDERS FORWARD.
QUICK MARCH.

Halt, Front. Dress.

1. The caution being given to FORM SQUARE, the colours fall to the rear, and the centre companies close by word of command from the senior company officer, to the centre.

The remaining companies are then faced to the right about, and on the word QUICK MARCH, bring up their respective right and left shoulders and move into square. Nos. 1 and 2 form the right face, Nos. 5 and 6 the left face, and the grenadier and light infantry the rear face of the square, being halted and fronted by their respective officers when they arrive in position.

2. *When*

Fig. 29—B.
THE SQUARE WILL MARCH TO FRONT, REAR, RIGHT, OR LEFT FACE.

Q. MARCH.

HALT, RE-FORM, SQUARE.

Q. MARCH.

2. *When the square or oblong is to march by any one face.*—The face which is to lead is announced; the colours move up behind its centre: The commander will then give the word to the two flank faces. BY SUB-DIVISIONS ON THE RIGHT AND LEFT BACKWARDS WHEEL; upon the word QUICK MARCH they wheel back, and the rear face advances two paces, and then faces about. The square marches two faces in line, and by their centre; and two faces in open column, which cover and dress to the flanks on which they wheeled back, carefully preserving their distances.—The square halts, and when ordered to RE-FORM SQUARE, the sub-divisions in column immediately wheel up, and form their faces, and the face which faced about, again faces outwards, and falls back the two paces it had advanced.

Should the baggage not admit of the sub-divisions wheeling back into the square, the flank sides may face by threes to the proper direction.

Fig. 29—C.

FORM LINE, RIGHT AND LEFT SHOULDERS FORWARD.
QUICK MARCH.

3. *To reduce the square.*—Upon the caution, the two centre companies open out by the side step to leave room for the colours, and the remaining companies receive the word RIGHT and LEFT SHOULDERS FORWARD, QUICK MARCH, and are brought into line by the echellon movement (*cc*), forming on the covering serjeants as usual.

Should the ground be uneven, or otherwise unfavourable to this mode of forming the square, the same operation may

 take

take place on the centre companies, by making the remaining companies of each wing form open column by the flank march of threes in their rear; the companies forming the side faces wheeling outwards, and the grenadier and light infantry closing to the front, and going to the right about.

All squares may be formed from a line of ten companies, in the manner instructed for eight companies, increasing the right and left faces accordingly; and a wing of five companies may form a square of its ten sub-divisions.

S. 91. *When a Battalion halted, in line, is to change its front to the rear, upon the centre.*

CHANGE FRONT TO THE REAR UPON THE CENTRE.

P. 26.

Upon the caution, the colours will countermarch; the two covering serjeants of the two centre companies change places, and will take up the points of formation (*ss*) to the new front, covering upon a centre serjeant, (Vide S. 73.) The leaders of these companies will then face them outwards, and will countermarch to the right, and change places by files.

THREES OUTWARDS. Q. or D. MARCH.

The commander, in the meanwhile, will give the word to the other companies THREES OUTWARD; and, upon the word QUICK or DOUBLE MARCH, the companies of both wings will countermarch to the right and left, and will pass each other by the left, those of the right wing describing a circle in their countermarch, the depth of a sub-division, to leave room for the others to pass. Each division will be halted in its

Front, turn, Halt, Dress up. — its relative position (*d d*) to the new front by the words *Front, turn, halt, dress up,* from its leader; and will form upon the coverers, who will run out as usual.

Fig. 26. In this mode of changing front to the rear by a movement along the former front, a fire may be at once opened from the new front (*c c*), and may commence when the centre companies (*s s*) have countermarched and be gradually augmented, until the whole is formed. (Vide S. 80, No. 5.)

CHANGES OF POSITION OF THE BATTALION FROM LINE, BY MOVEMENTS OF THE OPEN COLUMN.

(Plate V.)

ON A FIXED POINT.

S. 92. *When the Battalion is required to change position to the front on the right halted Company, by throwing forward the whole left, and by the flank march of Companies by Threes.*

Fig. 31. The right flank (*c*) is the fixed point on which the change is made, and is in the intersection of both lines. Another point (*b*) is placed, *ad libitum,* 20 or 30 paces beyond that flank. These two points determine the direction of the new line. If the change of front is to be at right angles (*ac*), with the old line, the right company stands fast; if oblique to it (*o c*), that company will wheel backwards on the right flank so many paces as will make it perpendicular (*c c*) to the new direction (*c o*). The

REMAINING COMPANIES THREES LEFT.

The remaining companies face by threes to the left, and disengage to the right.

FORM OPEN COLUMN IN FRONT OF THE RIGHT COMPANY. RIGHT COUNTERMARCH, Q. *or* D. MARCH. *Halt, Front, Dress.* RIGHT AND LEFT WHEEL INTO LINE.

The whole step off at the pace ordered, and the covering serjeants run out when at the usual distance from the new alignement, and take up the line of pivots from the right flank of the halted company (vide S. 74. No. 10.) The covering serjeant of the halted company will mark where the wheeling flank of that company is to rest (*d*) on forming line; and the covering serjeant of the company next to it, and which is to face the right halted one, will take distance (*ed*) from that serjeant: The companies are formed on their serjeants, and halted and fronted by their respective leaders, facing the halted company, and the whole will wheel up to the right and left into line.—(Vide S.76, No. 7.)

S. 93. When the Battalion is to change position to the rear on the right halted Company, by throwing back the whole left, and by the flank march of Companies by Threes.

Fig. 32.

The direction of the line is taken, as in the last section, upon the left flank (*a*) of the right company. If the change of front is to be at right angles (*a c*) with the old line, the right company will stand fast: If oblique to it (*b s*) the right company will wheel

wheel backwards on the left flank so many paces (*e g*) as will make it perpendicular (*a g*) to the new direction (*b s*).

REMAINING COMPANIES THREES RIGHT.
Q. *or* D. MARCH.
Halt, Front, Dress.
LEFT WHEEL INTO LINE.

The remaining companies face to the right, and disengage to the right. The whole step off at the pace ordered, and the covering serjeants run out, as before, and take up the line of pivots on the left flank of the halted company: Each company will form on its serjeant, and the whole will wheel to the left into line.

S. 94. *When the Battalion is to change position on a central halted Company, by the flank march of Companies by Threes, and that one wing is thrown forward and the other backward.*

REMAINING COMPANIES THREES RIGHT.
RIGHT WING WILL COUNTERMARCH.
Q. *or* D. MARCH.
Halt, Front, Dress.
LEFT WHEEL INTO LINE.

Fig. 33.

1. *If the line is to change its front to the left flank*—the points of direction will be taken on the left flank (*a*) of the central company: That company will stand fast, or will wheel backwards on the left (as explained in the last sections,) so many paces as may make it perpendicular to an oblique direction: The wings will face to the right by threes, and the left wing will disengage to the right, and the right wing to the left. They will step off at the pace ordered; the right wing countermarching by companies to the left, and the covering serjeants will run out, as before, and will take up the line of pivots from the left flank of

of the halted company; the covering serjeant of which will mark its wheeling flank as before; the left company of the right wing taking distance from that serjeant; each company will form on its serjeant, and the whole will wheel to the right and left into line.

REMAINING COMPANIES THREES LEFT, THE LEFT WING WILL COUNTERMARCH, Q. *or* D. MARCH. *Halt, Front, Dress.* RIGHT WHEEL INTO LINE.

2. *If the line is to change its front to the right flank*—the points of direction will be taken on the right flank (*b*) of the halted central company, which will either stand fast or wheel backwards on the right flank, in order to be perpendicular to an oblique direction, as already explained. The wings will face to the left, and the right wing will disengage to the left, and the left wing to the right, and form, as before, upon the covering serjeants in the line of pivots, and wheel up to the right and left into line (*c d*).

In difficult ground, or at the discretion of the commanding officers, the change of position, by means of the open column, may be effected by the flank march of companies *four deep*, Vide S. 47.

S. 95.

ON A DISTANT POINT.

S. 95. ***When the Battalion formed in line is to change to a distant Position, either to its Front or Rear, by the flank march of all its Companies, and this Position is either Parallel or Oblique to the one it quits.***

THREES RIGHT (OR LEFT.) QUICK MARCH.

Fig. 34.

Halt, Front, Dress.

The battalion will face by threes, and will disengage to whichever hand the new position outflanks the old one, for to that hand will the column have to incline during the march. The whole will step off at the word QUICK MARCH, and the commander of the battalion will remain with the head company (c.), and by making it insensibly advance (*a*), or keep back, (*b*) will regulate the heads of all the others during the march; and when this leading company halts in the line, the others will form successively in the new direction, and stand in open column at their just wheeling distances.—When the head of the column is within 30 or 40 paces of the new line, (its direction being already prepared,) the serjeants will run out and mark the pivot flanks of their several companies.

In this manner the commander, who is himself with, and conducts the leading company, moves in the direction that best answers his views, and at once takes up any position and to any front that is necessary.—As circumstances change his intentions, he may at every instant vary, and direct them upon new points of march: the rear of the column always conforming (without the necessity of sending particular orders)

ders) to whatever alterations of direction the head may take; and the commander conducting that head so as to enable the rear to comply with its movements without hurry.

If the new alignement is fixed upon a rear or central company of the column, all in front of such company will countermarch, so as to cover in column (Vide S. 74, No. 7, and as explained in S. 94.)

S. 96. *When the Battalion formed in line changes Position by breaking into Open Column, marching up in Column to the Point where its Head is to remain, and entering the Line by the flank march of Companies.*

Fig. 35.

THREES RIGHT OR LEFT. QUICK, OR DOUBLE MARCH.

The pivot flank of the column being directed on the adjutant (*c.*) who marks the flank point in the new line, will halt when arrived within a few paces of him: a point of direction (*d.*) beyond the adjutant is also immediately ascertained.—The word THREES (RIGHT OR LEFT, as is necessary to conduct into the new line) is then given and executed by all the companies which form successively on their covering serjeants.

HALT.

When a battalion in open column, entered and marching on a straight line, is to form at a point where its front flank is to be placed, it will receive the word HALT when its leading division is at a wheeling distance (*ac*) short of that point.

S. 97

S. 97. *When the Battalion formed in line changes Position by breaking into Open Column—Marching up to the Point where its Rear is to rest—And entering the Line by the wheeling of its Companies.*

Fig. 36—A.

Shoulders Forward. Forward. HALT.

Besides the officer (*c*) who marks the point of entry, two advanced points of march must be given (*o o*).—The battalion then enters by wheels on the moveable pivot, and when its last division is at its point, it receives the word HALT, and pivots being corrected, the whole are ready to wheel up into line.

Fig. 36—B. When a battalion open column enters a new position where its rear flank is to be placed (*a*), and the wheels are made to the pivot hand, it receives the word HALT when its rear division (*ad*) has just completed its wheel into the new direction.—If the wheels are made to the reverse hand, it receives the word HALT when the last division but one (*b*) has completed its wheel into the new direction, and the last division itself (*e* e) files and places its pivot flank at the given point.—When a battalion open column, entered and marching on a straight line, is to form at a point where its rear flank is to be placed, it will receive the word HALT when the pivot of its rear division arrives at that point.

Fig. 36—A.

S. 98.

S. 98. When a Battalion formed in Line changes Position by breaking into Open Column—Marching up in Column, and entering the new Position at an intermediate Point where a Central or any other Division is to rest and form in Line.

Fig. 37.

When the head of a battalion column by wheeling, enters the alignement at a middle, or any other intermediate point (*a*) where a central or other given division is to rest, the rear divisions will be brought into the new alignement by the flank march of companies by threes.

Shoulders Forward.
Forward.
HALT.
THREES RIGHT (OR LEFT.)
QUICK MARCH.
Halt, Front, Dress.

The leading company of the battalion having wheeled into the alignement followed by the others, and the named company being arrived at the point (*a*) where it is to form in line, the column is halted. The rear companies are then faced by threes to the right or left, and respectively conducted into the new alignement and formed upon their covering serjeants as above (*ccc*).

—

S. 99.

S. 99. *When the Battalion formed in Line changes Position, by breaking into Open Column—Marching in Column to the Point in the new Position where its Head is to rest, and to which its Rear Divisions form, by successively passing each other and wheeling up.*

Fig. 38.

Right or Left. Wheel.

Halt, Dress, March.

The column having arrived in any direction behind the line, and at the point where its head is to rest (*a*,) but which its rear is to pass, its leading division (1) will wheel into the line, and halt; each other division continuing its march will move on square behind the first formed division, at which point its leading officer will, if necessary, shift to its inward flank, and each, as it comes opposite to its ground, will successively wheel, upon a fixed or moveable pivot, and dress in line with those already in it.

If the column is marching in the direction of the line, it will of course, have its pivot flank on it, but as in this formation the wheel is made to the reverse hand, the battalion must, before it commences, shift the breadth of the column to bring the reverse flanks on the line, and be directed by them, the leading officers shifting at the same time.

In this manner the battalion does not stand in open column on the new line, but successively wheels up by divisions, and forms in full front on the given objects.—It may be used when the direction of its march is nearly in the prolongation of the new line, and when a battalion, arriving on the flank of a line already formed (*ca*), has to lengthen out that line.

OPEN

OPEN COLUMN MOVEMENTS.

(Plate VI.)

It is to be observed that all changes of position of the open column, made on fixed or distant points, are the same as those explained for the line, after that line has been broken into column; and therefore need not be repeated.

S. 100. *When the leading Flank of the Column is changed by the successive March of Divisions from the Rear to the Front.*

Fig. 39—A.

HALT.
BY SUCCESSIVE DIVISIONS REAR WING TO THE FRONT.
Threes Right, Front turn.

If the right is in front, the left to be brought up, and the column to continue to advance —The whole is ordered to HALT.—At the caution by SUCCESSIVE DIVISIONS REAR WING TO THE FRONT the officer of the left (the rear) company immediately orders it, *Threes right,—Quick March,* till his left flank can freely pass near the right flank of the others.—He then commands *Front turn,* close by the right flank of the company then preceding him.

Threes Right. Quick March. Front turn.

The officer commanding that company, as soon as the other approaches him, orders, *Threes Right—Quick March,* behind the now leading one, and *Front turn,* as before, when his inner flank is clear of the right flank of the former column, and he will thus follow the leading division, when at the due wheeling distance.—All the other

other companies successively perform the same operation; and when the right company has taken its place in the rear, the whole column is in perfect order.

Fig. 39—B.

If before this operation the column should be closed to half or quarter distance, then all the companies may be FACED at the same time, proceed as above directed, and each takes its distance from its preceding one, before it moves on.

S. 101. *When it is required to change the Wings of an open or half-distance Column, formed upon a Road where the space does not admit of the flank movement.*

Fig. 40.
BY DOUBLE FILES FROM THE CENTRE. REAR WING TO THE FRONT.
Inwards face by files from the centre, Quick March.

TWO CENTRE SECTIONS, OUTWARDS WHEEL, QUICK MARCH.
Halt, Dress.
Form company, Forward.
Inwards face, Quick March.

Form company, Forward.

Upon the caution being given, the rear company will receive from its leader the words *Inwards face—By files from the centre—Quick March,* and the sub-divisions will immediately wheel by files to the right and left (*a*): at the same time the commander of the battalion will give the words, to the remaining companies, TWO CENTRE SECTIONS—OUTWARDS WHEEL—QUICK MARCH, and the rear company will advance through the centre (*b d*) of the battalion by the opening formed by the wheel of sections: When the rear company shall clear the centre of the front company of the column (*o*), its leader will give the words *Form company—Forward;* each company will successively follow in the same manner, and the moment that the flanks of the

sections

sections which wheeled outwards are cleared by the company which is to precede from the rear, the words *Inwards face—Quick March*, will be given, as before,—the files of each sub-division inclining towards each other as they advance (*d*): and the whole will successively form to he front and take up the march in column (*c*).

S. 102. *When a double Column of Sub-divisions, formed on the centre, is required to form quarter distance Column of Companies right in front.*

Fig. 41—A.
THE RIGHT WING WILL FORM CLOSE COLUMN IN FRONT OF THE LEFT WING.
THE LEFT WING WILL CLOSE TO THE FRONT.
QUICK MARCH.

1. On the word QUICK MARCH, the left wing subdivisions close to the front, and the right wing sub-divisions move on, receiving the word *Left, Turn*, from their respective officers, as soon as their inner flank is open, and filing successively, are halted and fronted in front of the left wing. The battalion will then stand a single close column of sub-divisions right in front (*a c*).

FORM COLUMN OF COMPANIES.
LEFT SUB-DIVISIONS LEFT FACE.
QUICK MARCH.
Halt, Front, Dress up

The column of sub-divisions may then form column of companies by the left sub-divisions deploying on the right sub-divisions, upon which they will be *halted* and *dressed* by their respective leaders; and when this deployment is completed, the column will be formed at quarter distance and ready for any ulterior movements (GL).

In

In the same manner the double column may be formed into a single column *left* in front by the left wing sub-divisions filing successively, and forming in front of the right wing; and the deployments to form companies taking place as before.

Fig. 41—B.
RE-FORM DOUBLE COLUMN OF SUB-DIVISIONS, LEFT SUB-DIVISIONS RIGHT FACE, QUICK MARCH.
Halt, Front.

2. *Should it be required to re-form the column, as before,* the column will double up into column of sub-divisions by word from the commander, and each sub-division will be *halted* and *fronted* by the company leader.

THE RIGHT WING WILL COUNTER-MARCH,
Right Face,
Quick March,
Rear, Turn,
Halt, Front.

The right wing will countermarch by the front sub-division and the others in succession, receiving the word *Right Face, Quick March,* from their respective officers, and when the flank is open *Rear, Turn;* passing in this manner to the rear, and being halted and fronted when aligned with the corresponding sub-divisions of the left wing, thus forming a double column of subdivisions (*b d*).

S. 103. *When the Column, at open or half distance, is required to form a Square four deep.*

COLUMN WILL FORM SQUARE UPON THE FRONT COMPANY.
Q. *or* D. MARCH.

1. *If the Square is to be formed upon the front Company:* The caution being given, the remaining companies will advance upon the word QUICK or DOUBLE MARCH, and No. 1 battalion company will close upon the front company; and when No. 2, 3, 4, and

Sections outwards.

Right about face.

and 5 (supposing the right in front) shall successively gain the wheeling distance of sections, the leader of each will give the word *Sections outwards;* the rear sections immediately closing to the front after the wheel. The two rear companies, No. 6, and the Light Infantry, close upon the flanks of the sections, and come to the *right about,* by word from their respective leaders.

Fi. 42

FORM SQUARE ON THE RIGHT (OR LEFT) CENTRE COMPANY. RIGHT (OR LEFT) WING. RIGHT ABOUT FACE. Q. *or* D. MARCH. *Sections outwards. Front turn. Sections outwards.*

2. *If upon a centre company,* the caution will state on which centre company the formation is to be made, and that company will stand fast(*c*): The right wing (supposing the right in front) will receive the word RIGHT ABOUT FACE from the commander, and upon the word QUICK or DOUBLE MARCH, the officer commanding the named company, will wheel it outwards by sections, the rear sections closing on the front sections: the left wing will at the same moment close to section distance, and the companies will wheel in succession by *Sections outwards,* in a similar manner; while the right wing will move to the centre, with the rear rank in front; and when each company shall close up to the one which precedes it, the leader will give the word *Front turn—Sections outwards,* and the rear sections will close to the front, as before.

FORM SQUARE ON THE CENTRE.

3. *If an open column be attacked by cavalry while moving to the front or rear by the flank march of companies by threes—* the

INWARDS TURN.
Double March.

Sections outwards.

the caution will be given—and, upon the word INWARDS TURN, the divisions will turn to the right and left, at the same time forming two deep to their front and rear: and the instant the left centre company shall turn, supposing the right of the column in front, its leader will give the word *Sections outwards*—the wings will close upon the centre at the double march, and each company will successively wheel outwards into square, as explained in the last paragraph.

If the open column has its left in front, and is moving by threes to a flank, the square must be formed on the right centre company: The word INWARDS TURN is thus properly applied to the wings which turn to the colours.

When the attack has been repelled, the Square will form Column, as instructed in S. 107; open out again from the centre, and resume the flank march as may be required.

Fig. 43.
FORM THREE DEEP, MARCH.
FORM GRAND DIVISION SQUARES.

Q. or D. MARCH.

4. *A column of companies may form grand division squares.*—The whole will form THREE DEEP; and upon the caution to FORM GRAND DIVISION SQUARES, the leaders of the right companies (*a a*), will give the caution *by sub-divisions from the centre four paces backwards wheel*: Upon which the covering serjeants will mark four paces from the outer flanks: Upon the word QUICK or DOUBLE MARCH from the commander, the right companies will accordingly wheel backwards by sub-divisions on their serjeants, being *halted* by their respective leaders; while the left companies (*bb*) advance at the same time to half distance(*d*) and

and then receive from their leaders the words *right and left shoulders forward,* upon which the sub-divisions wheel up four paces to the right and left from the centre; and when the wheel is performed, they receive from the company leader the words *halt, right about face.* The sections of threes will close to the pivot flanks of sub-divisions during the wheel; and the different squares will thus be formed to flank each other at right angles (*c c c*).

RE-FORM COLUMN. Q. or D. MARCH.

5. The caution is given, and on the word QUICK MARCH, the right and left companies will *wheel up by sub-divisions* from their centre; the left companies advancing to full distance, where they will be *halted* and *fronted* by their respective leaders; and the sections of threes opening out to two deep during the movement.

When a battalion has been made perfect in this formation as above instructed, the squares may be formed at once upon the word *double march,* after the caution; the left companies forming threes on the march, and wheeling up by sub-divisions into square *right and left shoulders forward;* and the right companies forming threes during the wheel backward by sub-division from the centre.

FORM COLUMN OF GRAND DIVISIONS.

6. *A battalion column of companies may in the same manner form two squares of its respective wings.* The battalion will form grand divisions by the diagonal march, (Vide S. 37), the commander will then FORM THREE DEEP as before, and upon

FORM SQUARE OF WINGS.

the caution to FORM SQUARES OF WINGS the

the leaders of the right grand divisions will order, *by companies from the centre four paces backwards wheel;* the covering serjeants will take up the distance, to the rear from the eighth file; and upon the word

Q. *or* D. MARCH. QUICK or DOUBLE MARCH from the commander they will wheel backward, while the left grand divisions advance to half distance, and will then be *wheeled forward* by companies into square (as in No. 4), *halted* and *faced about* by their respective leaders.

The squares will be reduced to column of grand divisions, as already explained in No. 5, and column of companies will be reformed by the diagonal march.

These squares being three deep, the front rank only will kneel on preparing to resist cavalry.

QUARTER DISTANCE AND CLOSE COLUMN.

(*Plate VI.*)

S. 104. *When a Battalion forms a close or quarter distance Column from Line.*

Fig. 44. 1. *If before or behind either of the flank companies:* The caution will state which company, and whether the formation is in front or rear of it, and the officer of the named company (1) shifts, if necessary, to the flank which is to become the pivot one of the column: The covering serjeant of that company places himself six paces in front or rear of that officer, (according to circumstances

THREES RIGHT (OR LEFT).

circumstances) to mark the perpendicular of the front of the column: The other officers, if not already there, shift to the flanks which are to lead. The battalion will then face by threes to that company, and the heads of the other companies will disengage to the named flank.

Q. TO D. MARCH.

The whole will move at the QUICK or DOUBLE MARCH to the front or rear of the named company; covering serjeants stepping out when within six paces to take up the distances at the proper pivot point, and each leader will proceed in the same manner as in forming an open column from line, stopping in his own person at his covering serjeant, and giving his words *Halt, Front, Dress,* when the pivot flank of his company shall reach the covering serjeant, whom he immediately replaces; remaining himself perfectly steady, and giving his whole attention to the covering in column.

Halt, Front, Dress.

During the formation of all such columns, as soon as the battalion is put in motion, a field officer will immediately place himself in front of the column, before the serjeant of the named company, and from thence judging the perpendicular of the column, will attend to the serjeants covering each other in that direction as they come up, whether such covering is taken from the front or from the rear, which will depend on the formation of the column.

In close columns the supernumerary officers and serjeants will take post on the reverse flank. In quarter distance they will take post in the rear of their respective divisions.—When the column is right in front, the colours to be on the right of the

the left centre company, and when left in front, on the left of the right centre company.

Fig. 46.

THREES INWARDS.

Q. *or* D. MARCH.

Halt, Front, Dress.

2. *On a Central Company.*—Upon the CAUTION of formation being given, the officer of the named company will place himself on his future pivot flank, the other officers will shift, if necessary, to the flank which is to lead; the battalion will then face by THREES INWARDS; the heads of companies will disengage, according as they are to be in front or rear; and at the word QUICK or DOUBLE MARCH the rest of the formation will proceed as before directed, part of the battalion arranging itself before, and part behind the given company, and the officers covering on the proper pivot flank.

Fig. 45.

THREES RIGHT (OR LEFT.)

RIGHT OR LEFT COUNTER-MARCH,

Q. *or* D. MARCH.

3. *In forming columns facing to the rear,* the same operations take place, as to the front, except that upon the caution the named division (*a*) will *countermarch,* by command of its own leader, by files, and that the other divisions of the battalion FACE OUTWARDS from it, and will disengage to the right or left, according to which flank is to be in front, and will countermarch into column.

4. In the same manner column may be formed from line upon a central company facing to the rear; that company upon the caution countermarching by files—and the wings facing THREES OUTWARDS, and countermarching to the right or left, as before.

5. *The*

CLOSE TO THE FRONT.
Q. *or* D. MARCH.
or WITHOUT HALTING,
CLOSE TO THE FRONT.
DBL. MARCH.

5. *The close column may be formed from any more open column:* If on the march, by the head division *halting* upon the caution, and letting the others close up, and *Halt* successively.—Or, WITHOUT HALTING, by the rear divisions being ordered to close by the DOUBLE MARCH, and successively to resume the quick march; the front division having steadily continued that pace.

S. 105. *When the Column at close or quarter distance marches to a Flank.*

COLUMN WILL MARCH TO THE RIGHT (or LEFT) FLANK.

THREES RIGHT (or LEFT).
QUICK MARCH.

HALT, FRONT.

The caution will express to which flank the column is to march. The whole will then face by threes, and the officer who leads the front company, will take care to march in the exact alignement, upon the point that will be given: all the others will dress and move by him; and the leaders of divisions will be careful to preserve their proper distance from him. When the column HALTS, FRONTS, the pivot officers, serjeants, &c., shift to their proper places (if not already there) by the rear of their divisions.

S 106.

S. 106. *When the Column at quarter-distance moving to the Front, or Rear, takes ground to the Right, or Left, by the Echellon March of Sections.*

SECTIONS RIGHT (or LEFT).

FORWARD.

1. The command will be given that the column will move BY SECTIONS RIGHT (or LEFT), and the companies will form echellon by sections, as instructed in S. 65, taking the word FORWARD from the commander of the battalion, when the requisite wheel shall be made.

RE-FORM COLUMN.

FORWARD.

2. Upon the word to RE-FORM COLUMN, the echellon will wheel back by sections, as instructed in S. 65, the commander giving the word FORWARD when he sees the sections are in the line of their companies.

In this movement the officers commanding companies will remain on their proper pivot flank; and when the column takes ground to the reverse flank, the covering will be taken up on the reverse flanks of companies by the senior supernumerary officer, so as to preserve the exact parallelism of the companies when wheeled back into column.—Ground may also be taken to a flank by the diagonal march *Right (or left) half turn*, and this may be applied at close as well as quarter distance.

S. 107. *When the Column halted at quarter-distance forms a Square.*

Fig. 47.
FORM SQUARE.

1. Upon the caution being given, the pivot men of sections will face outwards—two to the right, and two to the left—and upon

QUICK MARCH.

Fig. 47.

upon the word QUICK MARCH, the front company will fall back upon the company in its rear, and Nos. 2, 3, 4, and 5, will wheel outwards by sections, the rear sections closing to the front after the wheel; No. 6, and the Light Company, will close up, and form the rear face of the square, receiving from their respective leaders the words *Halt, Right about face.*

Halt.
Right about face.

If the column should be on the march, and it be required to form a square, the commander will see that the quarter distances are corrected. He will then HALT, and form square, as above instructed.

PREPARE FOR CAVALRY.
READY.

2. *When the square is to resist cavalry*—The caution will be given, and upon the word READY, the two front ranks, all round the square, come to the kneeling position, slanting their firelocks forward, and resting the butts upon the ground: but the two kneeling ranks will not cock until required to fire. The standing ranks will fire by files; the firing to commence, and to cease, by signal from drum.

KNEELING RANKS,
READY,
P'SENT,
FIRE.
LOAD.

The kneeling ranks will fire a volley, when required by word of command from the commanding officer, either by faces or the whole at once. After the fire, the ranks will resume the posture of defence, and will come to the standing position, on the word LOAD.

The kneeling ranks may fire previously, resuming their posture

posture of defence while the standing ranks fire, according to circumstances, and the discretion of the commander.

RE-FORM COLUMN, QUICK MARCH. *Halt, Dress.* *Halt, Dress.* *Halt, Front, Dress.*

3. *To reduce the square.*—Upon the word RE-FORM COLUMN—the rear sections of the two side faces step back to wheeling distance, the pivot men of sections face inwards, and, upon the word QUICK MARCH, they will wheel backwards, and will be halted and dressed by the respective leaders of companies; at the same time, the front and two rear companies advance, and retire, and receive respectively from their leaders, the words *Halt, Dress,* and *Halt, Front, Dress,* when at the regular quarter distance. The square is thus reduced to the proper order in column, and the officers take post with their respective companies.

S. 108. *If a Battalion in close Column be attacked by Cavalry, it may form a solid Square.*

COLUMN WILL FORM SOLID SQUARE. OUTWARDS FORM FOUR DEEP. QUICK MARCH. Fig. 48.

1. On the caution being given, the officers and serjeants will face outwards, and proceed to the right and left into the intervals, relatively next to them, between the four centre companies. The centre companies will then receive the word OUTWARDS FORM FOUR DEEP; and upon the word MARCH, their sub-divisions form four deep to the right and left—the ranks of fours closing to the front as they form; while the two front companies will fall back

back upon the company in their rear; and the two rear companies will close up to the company in their front, and go to the right about by command of their respective leaders. The files at the angles will make a half face outwards, and the square is then complete.

The square will prepare to resist cavalry as instructed in the last section.

RE-FORM CLOSE COLUMN.

QUICK MARCH.

2. The caution will be given to RE-FORM COLUMN, upon which the ranks of fours of the centre companies will open out from the front to their proper distances; and upon the word QUICK MARCH, they will front form two deep. The two front companies will advance one and two paces; and the two rear companies will retire one and two paces, and front; the officers and serjeants will take post with their respective companies.

S. 109. *When a Column, halted at close or quarter distance, is to wheel on a fixed or a moveable Pivot.*

COLUMN TO THE RIGHT (or LEFT) WHEEL.

Fig. 49.

1. *At close distance on a fixed pivot.*—Upon the caution (supposing the right in front) the flank file on the left of the front company, whether officer or man (*a*), will face to the left; the front company stands fast, but the remaining companies make a half face to the right; the covering serjeant

Q. *or* D. MARCH.

jeant of the front company moving out to mark where the outward flank of the column will rest when the wheel is complete (*c*); Upon the word QUICK MARCH, the front Company will wheel as usual, with the exception that the wheeling-step must be shorter, to give the rear companies time to come round. These companies will step off at the same moment, bringing the right shoulder gradually up, and each file circling round and covering the relative files of the division in front: (Vide *S.* 77, No. 6.) The officers and covering serjeants will circle in the same way round the officers in their front, and confine their attention to the covering.

Halt.
Dress.

STEADY.

The leader of the front company will give the word *halt* when the wheel is complete; and each other company as it circles round will halt of itself, as the files cover the halted files of the division in front; and the word STEADY will be given by the commanding-officer the moment he sees that the rear divisions cover; after which no man is to move until companies are ordered to be dressed.

RIGHT (or LEFT) SHOULDERS FORWARD.

2. *At close distance on the moveable pivot.*—The column while in march may change its direction on the moveable pivot upon the same principle. Upon the word RIGHT (or LEFT) SHOULDERS FORWARD, all the divisions but the front one, (which will wheel according to the in-

structions

structions given in *S.* 23.) will make a half turn to the outward flank, and circle round, covering the relative files of the division in front; and the whole will continue to wheel until the word HALT, or FORWARD, be given.

HALT, or FORWARD.

3. *At quarter distance on a fixed pivot.*—Upon the caution, the pivot man of the leading division upon whom the wheel is made will advance six paces, and will then halt and face. The rear divisions half face to the reverse flank, as in the last Number, and upon the word QUICK or DOUBLE MARCH the front division will advance and wheel round the pivot, receiving the word *halt* from its leader as in No. 1; each succeeding division advancing and circling round to its relative distance in column, until the word STEADY is given by the commander (Vide S. 77, No. 7.)

RIGHT or LEFT WHEEL.

Q. *or* D. MARCH.

Halt.

STEADY.

4. *At quarter distance on a moveable pivot.*—The Commander gives the word RIGHT OR LEFT SHOULDERS FORWARD DOUBLE MARCH. Upon which the front division moves round at once to the new direction, and advances at a shortened pace, until the commander sees that the rear divisions have circled round into the alignement at proper distances. He then gives the word FORWARD—QUICK, and the quick march is resumed. (Vide S. 77, No. 7.)

RIGHT or LEFT SHOULDERS FORWARD. D. MARCH.

FORWARD. QUICK.

S. 110.

S. 110. *When a close or quarter-distance Column is to change its Front and Wings, by the wheel and countermarch of Sub-divisions round the Centre.*

Fig. 50—A.

1. Upon the caution being given, two points will be placed in the centre at (*a. b.*) to mark the angle on which the respective sub-divisions are to wheel the half circle (*c. c.*) and to mark the point at which the leading sub-divisions are to *halt*, when the countermarch is completed: The reverse sub-divisions will then receive the word RIGHT ABOUT FACE; and upon the word RIGHT WHEEL THE HALF CIRCLE, the rear sub-divisions make a half face to the left (supposing the right in front) as instructed in the last section: On the word Q. or D. MARCH the whole step off, and the right and left sub-divisions circle round, each occupying the ground which the other has quitted: The leading sub-divisions will *halt* when they shall respectively reach the points (*a* and *b*) on which the others had wheeled; and all the sub-divisions will *halt*, when they successively cover, as instructed in the last Section, and the commander, without pausing, gives the word FRONT DRESS. The word FRONT refers only to the reverse sub-divisions.

REVERSE SUB-DIVISIONS, RIGHT ABOUT FACE, RIGHT WHEEL HALF CIRCLE.

Q. *or* D. MARCH.

Halt.

FRONT DRESS.

2. *If the column at quarter distance is upon the march,* its front may be changed in a similar manner without halting, by the reverse sub-divisions receiving the words (supposing

RIGHT ABOUT TURN, LEFT SHOULDERS FORWARD.

Forward.

Front turn.

(supposing the right in front) RIGHT ABOUT TURN, LEFT SHOULDERS FORWARD, upon which the respective columns of sub-divisions wheel the half circle on the moveable pivot, their leaders giving the word *Forward* when the wheel is complete; and when the reverse sub-divisions get upon a line with the pivot sub-divisions (the one column advancing, and the other retiring) the leaders of the former give the words *Front turn,* which completes the countermarch; and the right sub-divisions immediately touch to the pivot flank of the column.

S. 111. *When a close or quarter-distance Column is to change its Front and Wings, by the wheel and counter march of Subdivisions, retaining the alignement of its front Division.*

Fig. 50—B.
REVERSE SUB-DIVISIONS RIGHT ABOUT FACE.
RIGHT WHEEL, HALF CIRCLE.
Q. *or* D. MARCH.

Halt.

1. Upon the caution being given, two points will be placed as before, at (*a. b.*) and the reverse sub-divisions will then receive the word RIGHT ABOUT FACE: the sub-divisions wheel to the right, as in the last Section; but when the pivot sub-division shall have completed the wheel of the half circle, it will *halt* at (*a*), and each of the other pivot sub-divisions will circle round and cover it in succession, and each will halt as it covers in column (*c e*). In the mean time the reverse sub-divisions will have wheeled the half circle round at (*b*), and

Halt.

FRONT, DRESS.

and will proceed along the ground which had been occupied by the pivot sub-divisions, and the leading division will *halt* when in line with the pivot sub-division; each succeeding division halting and aligning with the corresponding pivot division; the commander will then give the word FRONT, and the whole are formed upon the alignement of the former front.

FORM FOUR DEEP, MARCH. RIGHT (or LEFT) WING TO THE FRONT. *Quick March.*

Form two deep.

Fig. 51.

2. *A column formed at quarter distance may change its front and wings*—by FORMING FOUR DEEP, and the division from the rear passing through the intervals, upon the word *Quick March* from its leader; each division following in succession: When the leading division from the rear shall have passed through the intervals of the column, it will *form two deep*, and move on in column; and each successive division will do the same, when it shall have cleared the intervals: In the same manner a column may change its wings to the rear; and in both cases the leaders of divisions will place themselves in front of the pivot files, their covering serjeants in rear of them (*os*). (Vide *S*.80, No. 9.)

S. 112. *When a Column at close or quarter distance is to open out to full or half distance from the front or rear.*

1. *If from the front.* The caution will be given: The front company will stand fast; but the remaining companies, and the leader of the front company, will re-

RIGHT ABOUT FACE.

QUICK MARCH.

Halt, Front.

ceive the word RIGHT ABOUT FACE, upon which the covering serjeants will place themselves at the second file from the pivot flanks: the word QUICK MARCH will then be given, and the leader of the rear company will march upon an object, which will be taken by him, or upon a point placed for him. The leader of the front company will *Halt* and *Front*, the next company retiring from him when it shall reach the ordered distance—he fronting himself, as he gives the last word. The leader of each retiring company will halt when he receives the word, but he will not front with his company; he will remain steady to *halt* and *front* the next retiring company, and he will front himself with the last word. Thus the companies will successively open out; and if the leader of the rear company marches correctly upon an object, the covering will be accurately preserved.

In this mode of opening from the front, each leader will be answerable for the distance of the company which he *halts* and *fronts*.

QUICK MARCH.

Halt.

2. *If from the rear.*—Upon the caution being given, the covering serjeants move to the second file from the pivot flanks; the rear company stands fast, and the word QUICK MARCH is given to the remaining companies: The leader of the front company marches straight upon an object, and each leader from the rear will halt the company in his front at the ordered distance.

DEPLOYMENTS

DEPLOYMENTS.—*(Plate VI.)*

S. 113. *When the Battalion, in Column of Companies, at close or quarter distance (right in front), deploys into Line.*

DEPLOY ON THE FRONT COMPANY.

Fig: 52.

1. *On the front company.*—The caution being given, the base points will be taken up in the prolongation of this division, and corrected from it, just beyond where the left of the battalion is to extend (*d*).—The officer taking post on the right flank. The officers of the other divisions will change to the leading flanks of their companies.

REMAINING COMPANIES THREES LEFT.

Q. or D. MARCH.

Front turn.

Halt, Dress up.

At the word QUICK or DOUBLE MARCH, the rear companies step off with the heads dressed, moving parallel to the line of formation.

The officer of the second, or leading division, having stepped out to the right, at the above word DOUBLE MARCH, allows his division, led by his serjeant, to go on a space equal to his front, and then gives his word *Front turn.* His serjeant will run out, and take up the distance in the line, as directed in S. 74, No. 7, when one half the front of the company is clear of the division on which it is to form. He then steps forward and places himself before the left flank of the preceding division, giving the word *Halt, Dress up,* one pace in rear of the points of formation, and expeditiously corrects his men upon the serjeant, resuming his proper post in line.

In

Front turn.
Halt, Dress up.

In this manner every other company proceeds, each being successively fronted, marched up, and halted in line, by its officer, who stops on the left flank of the division which precedes him, and gives the word *Front turn* the moment his right flank is clear of it, so that there shall be no necessity, upon dressing up, to incline to the formed line.

DEPLOY ON THE REAR COMPANY.

Fig. 53.

2. *On the rear company.*—The caution being given that the line will be formed on the rear company, the officers commanding the remaining companies, and their covering serjeants, immediately pass behind their several companies, and post themselves on the right of each.

The base points will be taken up in prolongation of the rear company (*ss*), placing themselves in front of the flank files. (Vide *S.* 74, No. 7.)

REMAINING COMPANIES THREES RIGHT Q. *or* D. MARCH.

Rear turn.
Halt, front.
Dress up.

The whole, except the rear company, will step off at the word QUICK or DOUBLE MARCH, and will proceed to the rear by the word *Rear turn,* from each leader the moment his company is clear of the flank of that on which it is to form—the companies will then form to the rear in the same manner as to the front—passing the covering serjeant (who runs out as before,) and dressing up to the alignement. The leader having dressed his company from the right flank of the formed one, will take post on his right flank.

3. *On*

Fig. 54. 3. *On a central company.*—The same instructions apply to a deployment on a central company, with the exception that the battalion points (*bb*) must be given to both flanks in prolongation of such company, and that those of each wing face by threes outwards; one wing turning by successive companies to the front, and the other to the rear, and thus forming on the centre company.—(Vide S. 77, No. 9.)

Fig. 55.
DEPLOY ON THE REAR COMPANY ON THE FRONT BASE.
THREES RIGHT

Q. *or* D. MARCH.
Q. or D. March.
Halt, Dress.

4. *When a column deploys upon a rear company but upon the front base.*—The caution being given the base points (*s s*) will be taken up in prolongation of the front company, by the serjeants of the named rear one (6). The remaining companies will then face, and march as before; and the moment the front of the named rear company is clear, the leader will give the word *Quick* or *Double March*, and will *Halt Dress*, at his base points; each other company will successively *Halt, Front,* until it is uncovered, and will then advance on the alignement, observing all the details for deployment already given.

The same instructions apply to deployment on a central company upon a front base; observing that the companies in front of it will successively *halt, front,* (Vide S. 77, No. 10), and those in rear of it may *front, turn,* to the alignement without halting.

ECHELLON

ECHELLON FORMATIONS AND MOVEMENTS.

(Plate VII.)

S. 114. *When a Battalion from Line Wheels forward by Companies to either Flank into Echellon.*

COMPANIES —— PACES TO THE —— WHEEL. QUICK MARCH.

Fig. 56—A.

Pl. I. Fig. F.

1. At the general CAUTION, that the companies will wheel forward so many paces to the right or left, so as to place them perpendicular to their future lines of march, the officer, if not already there, moves to the named flank of his company, and the covering serjeant of each at the same time runs out, places himself at the outer shoulder of the 8th file from the named flank, immediately takes the said number of wheeling paces, on the circumference of the circle of which his flank man is the centre, and then stands fast with his body turned to the line of that flank man, who also faces into the line of his serjeant.—All those serjeants ought thus to be in a line (*ss*); but if any small correction is necessary, it will be made by the commanding officer, from the leading flank.

QUICK MARCH.

Halt, Dress.

At the word QUICK MARCH, each company wheels up till its 8th file arrives close behind the serjeant (*s*), at which time the officer who is on the standing flank gives his word *Halt, Dress*, eyes are turned towards him, and the dressing being completed, the serjeant places himself on the outward wheeling flank (*b*).

2. *When*

Fig. 56—B.

QUICK MARCH.

HALT.

Fig. 56.

2. *When the echellon thus formed marches forward and halts.*—The companies standing thus parallel to each other, and their leaders being on the pivot or flanks wheeled to, the whole move at the word QUICK MARCH, each flank on its own perpendicular: each officer must be attentive to preserve the distance he marched off at from his preceding pivot, and also his oblique covering in the line of pivots, as laid down in S. 78. These circumstances observed, the echellon may at any instant be ordered to HALT, and will then be in a situation ready to form parallel or oblique to the line it quitted. If parallel, by each division wheeling back to the flank of the one immediately behind it (B). If oblique, by the divisions moving up into the direction which the leading one then has (C), or may be placed in, as hereafter directed (D E).

S. 115. *When the Battalion, having wheeled from Line into Echellon, has marched, and halted, and is to form back parallel to the Line it quitted.*

WHEEL BACK INTO LINE.

QUICK MARCH.

Halt, Dress.

Fig. 56—B.

A CAUTION is given that the companies wheel back into line: on which the pivot men face into the line, and the officers take one step forward.—At the word MARCH each company wheels back to the pivot and on receiving from its officer the word *Halt, Dress,* eyes are turned towards him The line being completely formed, officers and

and serjeants (if not already there) move to their respective places in line, except in the occasional case of wheeling into line in the middle of a change of position: officers do not then shift from their leading flanks (unless ordered), but remain there ready to fire, and to resume the march in Echellon.

S. 116. *When the Battalion, having wheeled from Line into Echellon, has marched, and halted, and is to form up oblique to the Line it quitted.*

Fig. 56. Various circumstances attend the execution, according to the degree of wheel which must be given to the leading company, in order to place it in the required oblique position; and as the number of paces which have been already wheeled from line into Echellon determine the nature of the Echellon, they are an essential part of the following arrangements.

Fig 56—D. 1. *If the formation is made forward*, and the leading company is wheeled up the same number of paces (*be*) that it before wheeled from line into echellon (*ab*), then the others, without altering their situation, move on (ABD), and successively dress up with it. If the formation is made backward, the whole face about and form (Vide S. 118) when arrived in the new line. In this manner one or more battalions may change position on a flank or central company of the line.

2. If

Fig. 56—E. 2. If the wheel of the leading company exceed the number of paces (*ab*) which it before wheeled from line into echellon, the others wheel up one half (*e c*) of that excess (*e x*), move on, and successively dress up with it.

3. *If the formation is to be on the prolongation*

Fig. 56—C. of the front division as it stands, the others wheel back (*e o*) one half (*b d*) of what they originally wheeled forward (*ab*), then move on (BC), and dress up with it. These wheelings make the divisions stand perpendicular to the lines by which they must march to their points of formation, which lines (BC, BD, BE) change in consequence of the position given by the leading division.

S. 117. *When the Battalion formed in Line, changes Position to the Front, on a fixed flank Company, by throwing forward the rest of the Battalion.*

Fig. 57—B. When the commander has determined the new line to be taken, by placing a person (*a*) in it, 20 or 30 paces beyond the fixed flank: he orders the serjeant from before the 8th file of the flank company to wheel up into that line (*s*), thereby to ascertain the number of paces required.—He then directs that company to be wheeled and halted in the new position (*s o*), and the adjutant to prolong the line as far as the moving flank of the battalion will extend (B).

The

COMPANIES — PACES TO THE — WHEEL, QUICK MARCH. *Halt, Dress.*

The caution is then given to the other companies, to wheel towards it half the number of paces (*c*) wheeled by the flank one, for by that means will each stand perpendicular to the line (*ed*), which is drawn from its flank in the old line to its relative flank point in the new one, and it is along such line that each will move.—The battalion wheels into echellon, as in *S.* 114.

QUICK MARCH.

The officer being on the inner, and the serjeant on the outer flank of each company, the whole, except the fixed company, will move on at the word QUICK MARCH, as directed in *S.*114, No. 2.

—Shoulder forward.

Halt.

Dress up.

When each of the divisions in march respectively approach to 10 paces from the company already placed, the covering serjeant will run out to take up the distance (*d*) in the intended line, covering from the points on which the formation is made (Vide S. 74), and when the division leader shall reach the outer shoulder of the rear rank man, on the flank of the company formed before him, he will give the word *Shoulder Forward*; and when the division shall have wheeled up on the moveable pivot (Vide S. 23), exactly parallel to the intended line, with its front rank on a line with the rear rank of the formed company, he will give the word *halt*, proceeding nimbly to the formed flank in his front; and without unnecessary pause, giving the word *Dress up*, correcting his men upon the

the serjeant (*d*) and the battalion point (B). The serjeants will remain on the outward flank until replaced by the leader of each succeeding company.

The exact formation in this oblique line depends entirely on the companies having wheeled (only) one half of the angle which the new position makes with the old one; for should they at first wheel the whole of that angle
Fig. 57—C (*eo*), they would be then marching parallel to that line, and arrive in it doubled behind each other;
Fig. 57—B. whereas, by having the other half of the wheel to complete, when they come near to the new position, each moves in a perpendicular direction, and vacates the ground required by the succeeding one to form upon.

In the formation of oblique lines from the echellon march, it is quite essential that the divisions should not be wheeled up parallel to the intended line, until the front rank man on the leading flank of each division, shall be clear of the rear rank flank of the division formed before it; for if the wheel
Fig. 57—C. is made sooner, the flanks must more or less overlap; and crowding or shifting by an irregular movement, must ensue to gain the line: By the front rank of each echellon division halting in a line with the rear rank of the formed division, it will have one clear pace to dress up to the alignement.

S. 118. *When the Battalion changes Position to the Rear on a fixed flank Company, by throwing backward the rest of the Battalion.*

Fig. 58—B The new position is given (a d), and the flank company (*a*) wheeled into it in the manner already directed, but backwards instead of forwards.

The

RIGHT ABOUT FACE.
COMPANIES —— PACES TO THE —— WHEEL,
QUICK MARCH.
Halt, Dress.

The rest of the battalion FACES to the right about, the companies then wheel forward the given number of paces towards the standing flank — or they may wheel BACKWARD into Echellon, and then FACE about.

MARCH.

Fig. 58—B.

—*Shoulder forward.*
Halt, Front.

Dress up.

The companies MARCH with their rear ranks in front, the serjeants running out at the proper time as before, and will form in line in the same manner as when changing position forward; that is, when the officer of each shall see the proper front rank man of his leading flank (*b*) clear of the rear rank man of the flank (*c*), on which he is to form, he will give the word *Shoulder forward,* wheeling past the covering serjeant, (a file giving way) who had taken up the point of formation, and *halting* and *fronting* in the rear *(ee)*, with the front rank upon a line with the rear rank of the formed company as before; he remaining at the flank of appui, and giving, without unnecessary pause, the word *Dress up*, correcting his men upon the serjeant and battalion points.

S. 119. *When the Battalion changes Position on a Central Company, by advancing one Wing, and retiring the other.*

Fig. 58—A, B. 1. The company (*a*) of the wing to be thrown back is wheeled backwards, and the company of the

the wing to be brought forward, is wheeled forward; or a central company is wheeled upon its centre, into the new direction. Two points (d d) are quickly taken in the line, about where the flanks of the battalion are to extend, and in the line of the central company.

— WING, RIGHT ABOUT FACE. COMPANIES —— PACES INWARDS WHEEL. QUICK MARCH. *Halt, Dress.*

The retiring wing (B) FACES about—both wings wheel their companies inwards and forwards, half as many paces (*h*) as the central companies had wheeled.

MARCH.

The whole MARCH forward into line with the central companies, the advancing wing (A) dressing up, and the retiring wing (B) fronting and also dressing up (*ee*), as already directed in S. 117 and 118.

S. 120. *When from open Column, the Companies wheel backward into Echellon, in order to form in Line on the front Company.*

Fig. 59.

1. The front company either remains square to the column, or is wheeled backward into the intended direction of the line; and on the position given to it must depend the relative one which the other companies will take. If the front company (*oc*) remains square to the column, the remaining companies wheel backward on the reverse flank four paces, or the 8th of a circle (*e c*); but if the direction be oblique (*o b*),

(*o b*), then the remaining companies wheel one-half the number of paces (*cb*) wheeled by the front company, in *addition* to the 8th of a circle. Thus, if the front company wheels 2 paces, the remainder wheel 5; if 4 paces, they wheel 6, and so on.

COMPANIES — PACES, ON THE — BACKWARD WHEEL

Fig. 59.

On the CAUTION, that the companies, except the front one, will wheel back on the right or left so many paces, (and which wheel is always backwards, and always on the reverse flank of the column, as being that which afterwards first comes into line) the officer moves to that flank, and the serjeant of each places himself with his back to the outer shoulder of the 8th file of the rear rank, immediately takes his named paces (*s*), and halts, fronts, with his body turned in the line of the flank man on whom he wheeled. At the word QUICK MARCH, the company wheels back till the 8th file of the rear rank touches the breast of the serjeant, (who gives a low caution to halt,) it is then halted and dressed by the officer from the standing flank, the serjeant places himself on the outward flank, and the whole are now in a situation to march forward, and form in the line on the head company, as in S. 117.

QUICK MARCH.

Pl. I. Fig. F.

Halt, Dress.

Fig. 61—(3 G) 2. *If the line be formed on the rear company of the column*—that company (3) will stand fast; the others (2, 1, G.) will FACE ABOUT—wheel BACK on the pivot flanks of the column, as being those which afterwards

wards first coome into line—QUICK MARCH, and then *Halt, front*, successively in the line of the rear company (3).

3. *If the line be formed on the rear company, but facing to the rear*—the whole column will countermarch each company by files, and then proceed as in forming on a front company.

Fig. 61—AB. 4. *If the line be formed on a central company of the column*, that company will stand fast. Those in front of it will FACE about. The whole, except the central company, will WHEEL back the 8th of a circle or four paces: those in its front on the proper pivot flanks of the column, and those in its rear on the reverse flanks, such being the flanks that first arrive in line The whole would then QUICK MARCH into line with the central company, as in S. 117 and 118. If the line is to be formed facing to the rear of the column, the divisions will countermarch, and the new front will be thrown back and the rear forward as before.

Fig. 61—CD. 5. *But if an oblique line is to be formed on a central division*: The central company (3) will wheel backwards into the new direction on its reverse flank; and the companies in its rear will wheel backwards on their respective reverse flanks, a number of paces (*f g*), in proportion as laid down in No. 1: But in order to bring the companies in front of the central company into lines of march parallel to each other, with their eading flanks perpendicular to the points of of appui in the new line, and thereby to prevent crowding and false movement, it is necessary that the front companies should FACE ABOUT, wheel into line (*a b*) to their proper reverse flanks, and WHEEL FORWARD to their proper pivot flanks so many paces (*h i*) (keeping the rear rank in front) as, in addition to those wheeled by the rear companies, may complete the quarter circle; thus if the rear companies have wheeled six paces (*f g*) the front companies will wheel two (*h i*), if five, three, if seven, one; and the whole will then form in line as laid down in S. 117 and 118.

S. 121. *When an open Column is required to form Line to the rear on the rear Company, by the Echellon March, facing to the rear.*

Fig. 60.
FORM LINE TO THE REAR.
Left wheel, Half circle, Quick march.
LEFT WHEEL INTO LINE.
QUICK MARCH.
FOUR PACES, LEFT WHEEL.

1. *If the right is in front.*—Upon the caution, the rear company will wheel to the left the half circle (*a*), by command of its leader, and points will be taken up in its prolongation (*b*).

The remaining divisions will then wheel the quarter circle to the left into line (*c*), and then one-eighth of the circle (*c d*), which will bring their inner flanks upon perpendicular lines to the points of *appui* (*o o o*), in the new alignement; and they will form accordingly by the echellon march on the rear company.

2. *If*

2. *If the left is in front,* the same formation will be made to the right on the rear divisions: And if the column is in march, it may be performed by the rear division wheeling the half circle, and halting; the remaining divisions at the same time wheeling on the moveable pivot, until the commander shall see that they are square for the echellon movement on the halted division, when he will give the word FORWARD.

S. 122. *When from Line, the Companies of a Battalion march off in Echellon, successively and directly to the Front, and again form in Line, either to the Front, or to the Flank.*

Fig. 62—A.

1. *So long as the intention is to form to that front;* they may march off at any named distance whatever behind each other, and when the leading division *Halts* (*d*), the others (*b*, *&c.*) may move on, and dress in line with it (A).

Fig 62—B.

2. *But when it is the intention to form line to the flank:* If the echellon be formed at regular distances, the divisions of which it is composed, may wheel to the right or left (b) (according to which flank is in front) thus placing the rear division of the echellon (*b*) in advance, and the whole may successively form line (B) upon any named division (b) perpendicular to the former front:

Fig. 62—C

If from the direct echellon it be required to form an oblique line, the directing flank of the leading division (C) will be considered the first point in the intended line,

 and

and its precise direction will be established, by placing another point (*a*) beyond and before it. The covering serjeants will run out upon the caution, and take up the line of covering from the echellon flanks, at distances (*s s*) equal to the extent of the division which respectively precedes them. The rear companies will then be moved by the diagonal march (*c x*) upon their respective serjeants; or if the line is taken up more to the flank they will be conducted by the combined application of filing (*xs*), and the diagonal march (*c x*) to those serjeants, halted and fronted (*s s*) square to their former front; and the oblique line is formed by the wheel back of each company on the established echellon flanks; or when the companies are thus placed, the whole may be put in march to the front, HALT and WHEEL backward into line at some more advanced point.

3. *If a column is to be formed to the flank from a direct echellon,* the covering serjeants will of course, take up the proper pivot points from the leading division, and those in the rear will move by threes into the alignement.

S. 123. *When a Battalion halted, or moving to the Front in Echellon of Companies, is required to form Grand Division Squares, to resist Cavalry.*

RIGHT (or LEFT) COMPANIES HALT.
Halt, Dress.
Fig. 63.—A.

FORM GRAND DIVISION SQUARES.
Front and Rear form three deep.

QUICK MARCH.

Halt, Front, Dress.

1. If the echellon is in movement, the commander will HALT the right or left companies of grand divisions, according to which flank leads the echellon; and the leader of each grand division will form the rear company upon the one halted. Upon the command to FORM GRAND DIVISION SQUARES, each leader of grand divisions will give the word *Front and rear, form three deep, March,* upon which the left sub-divisions of the right companies (*a*) will form three deep to the front, and the remaining sub-divisions will form three deep to the rear: Upon the word QUICK MARCH the left sub-divisions of right companies close the threes to their centre, the right sub-divisions (*b*) move left shoulders forward, and form the right faces of the squares, and the left companies move by sub-divisions, right shoulders forward, and form the left and rear faces of the squares, each closing their files of threes on the march to the inward flanks. When these sub-divisions have wheeled round to form the flank and rear faces of the squares, they will receive the words *halt, front, dress,* from their respective officers, who move during the formation to the interior of the squares.

 2. If

Fig. 63—B.
FORM COLUMN OF GRAND DIVISIONS.
RIGHT COMPANIES RIGHT ABOUT THREE QUARTERS FACE.
QUICK MARCH.
FORM GRAND DIVISION SQUARES.
THREES LEFT.

2. *If the echellon of grand division squares is to be formed from open column.*—The commander will form column of grand divisions by the diagonal march, (Vide Part I. Sect. 37), and upon the word form GRAND DIVISION SQUARES, the leader of the front grand division will proceed as explained in last paragraph.

The other grand divisions (*dd*) will receiv the word THREES LEFT OF RIGHT, according to which flank is in front, and each leader will halt his grand division, when a sufficient number of paces to the flank of the one halted in his front, to throw the squares into their echellon positions, for mutual defence; the number of paces will depend upon the strength of the battalion.—Each leader will then quickly form grand division square, without fronting his division, by the word *Right* and *left face,* upon which, the left subdivision of the right company fronts and closes into its threes.—Upon the word *Quick March,* the others wheel round into square, as described in the last number.

Right and Left Face.

Quick March.

RE-FORM LINE OF GRAND DIVISIONS ON THE FRONT FACES.
Right and left shoulders forward,
Quick March.

3. *When the squares thus formed are required to re-form echellon of grand divisions on the front faces.*—Upon the command, the leaders of each grand division will give the word *Right and left shoulders forward, Quick March*; upon which the front faces will open out from the centre, and front form two deep The other sub-divisions will open out on the march, and will be brought into line by their respective officers,

cers, front formed two deep; the grand division leader dressing them from the leading flank of the echellon.

When the squares are reduced into their relative grand division lines, the column of grand divisions is reformed, as instructed in Section 122, No. 3; and column of companies will be formed by the diagonal march, (vide S. 37).

When the squares fire, the front rank of threes come to the kneeling position.

Should the Echellon of squares be required to move by any of the faces, it will be done as directed for the movement of the rallying square (S. 67.)

S. 124. *When a Battalion changes Position from Line by the Echellon movement of Sub-divisions or Sections.*

THREE PACES, RIGHT AND LEFT WHEEL. QUICK MARCH.

Halt, Dress. Fig. 64—A.

1. *If the battalion is halted* the CAUTION is given that the sub-divisions or sections will wheel 2 or 3 paces to right or left.—At the word QUICK MARCH, the outward man of each subdivision or section, whatever its strength may be, wheels up 3 paces, and each company officer gives the word *Halt, Dress:* the 3 file of colours and centre serjeants will wheel up as a separate division parallel to the others

QUICK MARCH. Fig. 64—AB.

At the general word QUICK MARCH, the whole move on in their then perpendicular direction, as specified in the echellon movements (B).

HALT.

The whole halt.

2. If

Fig. 64—B.
WHEEL BACK INTO LINE.
QUICK MARCH.
Halt, Dress.

2. *If the Battalion is to resume its former front* (*b c*) it receives the CAUTION to wheel back into line.—At the word QUICK MARCH, each division wheels back 2 or 3 paces, thereby joining the next standing pivot, and is halted and dressed by the leader of each company, as already explained.

FORM LINE.
QUICK MARCH.

3. *But if the Battalion is to form forward,* after the HALT, in the direction (*c d*) of its leading division, the other divisions will wheel back half the angle, which the leading division forms with the original line, or that division is previously wheeled up into a more advanced direction (*e c*)—The company leaders will shift, if necessary, each to the flank of his leading division or section. A CAUTION is given to form line, and at the word QUICK MARCH, the whole, except the head division, move on, and the line is formed by the successive echellon of sections, in the manner already laid down for companies; but as each section comes into the line, it is dressed by the officer commanding the company from the formed flank on the covering serjeant, and not by the supernumerary officers in the temporary command of it who fall to the rear.

4. *If the battalion is in march in line,* and the object is to gain ground to the flank by the echellon march of sub-divisions or sections,

SUB-DIVISION, OR SECTIONS. RIGHT OR LEFT.

sections, and without making a previous halt. On the word SUB-DIVISIONS or SECTIONS, RIGHT or LEFT, the pivot men of the front rank of each turning in a small degree to the pivot hand, mark the time for two or three paces, during which the named divisions wheel on those men, and the 3 files of the colours and centre serjeants also wheel up as a division, parallel to the others. At the third or fourth pace, and at the word FORWARD, the whole move on direct to the front which each division has acquired, the position of leaders being as already described.

FORWARD.

RE FORM LINE.

FORWARD. HALT.

5. When sufficient ground has been taken to the flank, on the word RE-FORM LINE, the pivot men mark the time for three paces, turning back in a small degree to their original front, and the sub-divisions or sections instantly wheel backward into line, without altering the time, and at the fourth pace the whole step on having received the word FORWARD, till the battalion is ordered to halt.

FORM LINE ON THE LEADING DIVISION. HALT.

6. When sufficient ground has been taken in echellon to a flank, and that a forward formation of the line is to be made, the head division is wheeled up 2 or 3 paces more and halted. The rest of the battalion receives a CAUTION to form on the head division, and proceeds as directed in *S.* 117.

END OF PART THIRD.

PART IV.

LIGHT INFANTRY.

(Plate VIII.)

GENERAL PRINCIPLES FOR LIGHT INFANTRY FORMATIONS.

S. 125. *Movements, &c.*

Object of light infantry movements.

1. The object of light infantry movements, whether in battalion or in companies, is to protect the advance or retreat, and to cover and assist the manœuvres and formation of larger bodies; and these particular instructions are laid down to simplify the service of light infantry, to establish uniformity of practice and of movement, and to afford such details fo the drill and formation, as will, under the most extended circumstances, contribute to produce unity of action and of result.

To be well versed in all movements of the line.

2. The first principle essential to this object, is the utmost rapidity of movement, consistent with order and regularity:—But as greater celerity has now been given to the movements of the whole line, the rates of march laid down in the foregoing parts, apply to light infantry battalions and companies; and it is required of every battalion of light infantry,

infantry, that it shall be thoroughly versed and well grounded in the prescribed exercise and movements of a battalion of the line; for no latitude can be permitted in the mode of executing the drill and manœuvres laid down in the first, second, and third parts of this book, upon the plea that they are performed by light infantry battalions; and such extended movements and formations as apply to this particular description of force, are laid down in *this part,* as applicable to them, when employed in extended formations.

Battalions of the line required to practise movements in extended order.

3. When battalions of the line are in perfect order in all the detail of line movements, it is essential that they should be practised in certain extended formations. It is always desirable that a battalion of the line, in the absence of any force of light infantry beyond the light companies of regiments, should be competent to assist in protecting the front and flanks of a column of march; and the formation of an advanced guard (Vide S. 130,) and the posting of piquets (Vide S. 132,) apply to all descriptions of infantry corps.

4. Light infantry companies must, upon the same principle, be perfectly versed in all the duties of the line, and ready at all times to act in their place in battalion.

Formation and telling off.

5. Light infantry battalions and companies, are formed and told off in the same manner as battalions and companies of the line. Except that Light Infantry battalions having no flank companies, are told off from right to left.

S. 126.

S. 126. *Signals and Sounds for regulating Movements.*

1. Signals and sounds are necessary in various situations; they are intended as substitutes for the voice: But as they are liable to be misunderstood, they should never be resorted to, excepting when the voice cannot reach; or for the purposes of drill and instruction.

To be as few and as simple as possible.

2. For this reason, and as the same sound upon a different key, or in different time, is apt to occasion mistakes, they ought to be as few and as simple as possible.

No movement to be executed until the last sound of the bugle.

3. No movement should ever be executed until the bugle sound is perfectly finished; and in the combinations of the sounds with the "*Fire,*" *that* sound should always be the *last*, otherwise the company might immediately commence a fire upon the spot; and if the march or retreat were to follow, it would not be heard.

4. The following sounds appear sufficient for every situation in which light infantry can be employed:—

The light infantry call and The officers call } as established, and therefore not numbered.

I. To Extend—from that part of the line where the bugle sounds, except preceded by the distinguishing G.

II. To Close—to the spot from whence it proceeds, and for skirmishers to run in upon the supports, except preceded by the distinguishing G.

3. To

III. To March—in order of the present formation.

IV. To Halt—in the same order; excepting in advancing or retiring from line by *files*, in which case they form up to the front.

V. To Fire.—If when halted, they fire upon the spot, skirmishers selecting their objects. If on the march, whether advancing or retiring, it will be by alternate *ranks*, if in single files; by alternate *files*, if in double files, unless otherwise ordered.

VI. Cease Firing.—Every man to cease firing and load.

VII. To Retreat.—When not firing to retire immediately in *quick* time. The line, reserves, and skirmishers, facing to the right about, if no other order or rate is specified.

VIII. Assembly.—This sound may be used on many occasions, *viz.*, to turn out a corps, or company, at any time by day or night: to repair to a place of rendezvous previously appointed, when extended as skirmishers, and surprised by cavalry in open ground. For skirmishers, with their supports and reserves, to run in upon the battalion.

IX. Disperse.—The whole to disperse according to the object and the orders given.

X. Skirmish.—To send out any portion to skirmish. This may be indicated by each sub-division and section (or each company if a battalion) having its distinguishing sound.

XI. Incline to the Right. *Left shoulders forward.*

XII. Incline to the Left. *Right shoulders*

ders forward.—Whether marching in close or extended order, this is obeyed by bringing forward the shoulder gradually.

XIII. The Alarm.

XIV. The Lie-down.

XV. The Rise.

5. The following signals being repetitions, or combinations of the preceding sounds, are not numbered:

To Annul.—Whenever the halt is sounded, it is considered as annulling every previous sound excepting the "*Fire;*" therefore, if the company or battalion be inclining to the right, or left, or extending in any direction, upon the halt being sounded, they are to stand fast, and the subsequent movements will depend upon the sounds that may hereafter be given, without any reference to the former sounds.

Forwards.—When the direction has been sufficiently altered after inclining to the right, or left, the bugle will sound the "March," which in this situation signifies "*Forwards.*"

Incline to the Right and Left—Nos. XI and XII. These two sounds immediately following, signify that a chain or line of skirmishers, an advanced or rear guard, should occupy more space to the right and left. The space to be occupied if no certain number of paces is fixed upon, should be one half of their original extension. When they have sufficiently increased their distances, the "March" will be sounded. Should the increased extension not be sufficient, the sound should be repeated. In increasing their distances

tances, they are to continue their front and other operations, should they either be firing or advancing, and extend themselves by degrees from the centre.

Distinction between the assembly and the close.

6. If when the battalion or company is detached, the skirmishers have to close, they always run in upon the supports in the *first* place, and the assembly will afterwards be the signal for the *whole* to close in upon the battalion:—But if the assembly sounds first, without any close, it is a signal for the whole to make the best of their way to the rear of the battalion; in which case they must move as rapidly as they can, as this implies the necessity of greater expedition. If skirmishers are pursued when the assembly is sounded, they should be taught to keep wide of the battalion.

7. The MARCH—RETREAT—HALT—FIRE—CEASE FIRING, and DISPERSE, are the only sounds which should be repeated by all the buglers on every occasion.

How the use of the Bugle may be increased.

8. The use of the bugle may be considerably increased by adopting the use of three simple G's, as distinguishing sounds.

One G to denote the right of the line

Two G's, the centre.

Three G's, the left.

This, preceding any sound, denotes the part of the line to which it applies. For instance, two G's before the EXTEND, signifies to extend from the centre. One G followed by the CLOSE, signifies to close to the right. When no G is prefixed to the EXTEND, it will

will mean from that part of the line where the bugle sounds.

9. There should be a pause of three seconds between all orders by sounds.

10. Signals by bugle sounds do not apply to bodies of troops in reserve.

How to regulate the movements of a company or battalion flanking at a distance.

11. The movements of a company or battalion at a distance, may be regulated by the bugle. When it sounds the double time, it indicates that the utmost expedition, consistent with good order, is necessary.

Time of movement.

12. When no particular time is specified, all light infantry movements in close order, excepting formations from file, are in quick time: All formations from file, and from extended order, and all extensions also are in double time. A just discretion, however, is necessarily vested in every commanding officer on actual service, when the double time must be sparingly used. In broken grounds, or when rushing in advance to seize an advantageous point, or in cases of great danger in retreating and in assembling, it may always be resorted to; but for common skirmishing it is liable to exhaust the men.

Situation of the light infantry companies in battalion.

13. The light infantry company will always occupy its place on the left of the battalion till called for. When the call sounds, the company will order arms, and unfix bayonets, without word of command, and will be ready to move.

S. 127. *Skirmishing.*

1. A battalion or company may extend its files from any part of the line, and at any distance, either by single or double files; but a battalion had better extend by double files, as it is more readily accomplished in that manner for a large body. If this particular mode, however, is not specified for a detachment or corps, it will extend by single files.

Detached skirmishers

2. Detached skirmishers must be governed by circumstances and situation; and fire kneeling, lying, &c., as either of these may require.

Number of paces in advancing or retiring.

3. The number of paces in advancing or retiring, must be regulated in the same manner, by the superior officer commanding.

General rule in advancing and retiring.

4. It is a general rule, that in advancing, the men advance by the right of the men in their front, and in retiring, by the left of the men in their rear.

How to proceed when no mode of skirmishing is specified.

5. When extended in *single* files, and *no mode of skirmishing is specified,* it will be by alternate ranks; when extended in double files, it will be by alternate files.

All preparatory cautions to be passed along the line before the bugle sounds.

6. All commands to which the bugle sounds do not directly apply, to be passed in the first instance distinctly along the rear, by the officers or non-commissioned officers. In many cases the following few signals by sword, which every soldier can easily comprehend, will supersede the necessity of others:

FORWARD.—The sword at arm's length pointed.

To THE RIGHT.—The sword at arm's length, to the right.

To THE LEFT.—The sword ditto, left arm to the left.

HALT.—The sword held up perpendicular.

Skirmishers to overlap the flanks of lines.

7. In covering the advance of lines, skirmishers will take care to protect and overlap the flanks.

How to occupy the edges of hills, &c.

8. In occupying the edges of hills, or the backs of fences, whether in close or extended order, the line will always follow their direction, provided the salient angles are not too acute; but the men must be very careful to fire clear of each other.

Relieving skirmishers.

9. In relieving a line of skirmishers, the new line extends in the rear, out of reach of the enemy's fire, and afterwards runs up rapidly to the old line; each file of the former proceeding straight in rear of the latter, so as to keep them between the enemy's fire.

When halted.

10. If the relief is to take place when halted, each file of the old skirmishers runs straight to the rear, the instant that a file of the new skirmishers reaches the line of defence; and whenever the former is out of reach of the enemy's fire, they close in upon their supports. Should an immediate advance be intended, the relieved skirmishers ought to remain in the line, if *covered*, instead of exposing themselves to a fire whilst retiring.

When advancing.

11. If the relief takes place while advancing, the new skirmishers will run up in the same way, and pass briskly in front of the others; the

the old skirmishers *lie down* till they are out of the enemy's fire, after which they close upon their supports as before.

When retiring.

12. If relieving while retiring, the new skirmishers extend a considerable distance in rear, and each man looks out for a good situation. The old skirmishers continue to retire in their usual order, until within 20 or 30 paces of the former; they then run through them to the rear, until they are out of reach of the enemy's fire, after which they close.

Supports and skirmishers to relieve each other.

13. Those which have been acting as supports may relieve their own skirmishers in this way,—in which case the latter afterwards form in as many parties of reserve as the others consisted of, closing to the right and left accordingly, when out of reach of the enemy's fire.

The whole relieved, or strengthened, or diminished.

14. But if the reserves and skirmishers are *all* relieved by fresh parties, each of the supports preserve the relative position with respect to *their own* skirmishers, until the two lines have relieved each other. Any *part* of a line of skirmishers may be relieved in the same manner:—It may also be strengthened by throwing forward one or more companies or sections to particular parts of the line; in that case they must mix with the others, and divide the distances, or it may be weakened by calling in one or more sections; in which case the remaining skirmishers will extend to the right and left, so as to cover the vacancies of those who retired.

DETAIL

DETAIL OF FORMATION.

S. 128. *To cover the Advance and Retreat of the Line.*

To extend from the right.

1. As soon as the order is given, (either by word of command, or by bugle), the officers drop to the rear:—The captain places himself in rear of the centre; the first lieutenant is attached to the front line of skirmishers; the second to the rear line; the third lieutenant remains with the captain, ready to receive his orders. The serjeants fall likewise to the rear,—but two remain posted directly behind the centre, (unless the company has been acting detached, in which case they are *in* the centre), ready to direct the lines in their advance. At the last sound of the bugle, the right-hand file stands fast; the remainder trail arms, face to the left, and extend.

When soldiers are drilled by word of command, they move or obey in the same manner, at the *last* word, which should be given short:—PACES—FROM THE RIGHT—EXTEND.

PACES—FROM THE CENTRE—EXTEND, &c.

TO THE RIGHT—CLOSE, &c.

ADVANCE—HALT—FIRE—RETIRE, &c.

Light companies should often be practised in judging their own distance of files; the points on which the flanks are to rest being previously notified.

2. There is a particular sound for double time, as above specified, which, if necessary, the men can assume upon the march; but

 the

the files must be loosened before they attempt it.

3. The front rank men of files move straight before them, covering correctly on the march; their respective rear rank men cast their eye over the right shoulder, and tap their front rank men, at the distance of two, four, six, or any other given number of paces, as a signal for them to halt and front.

Distance of files.

4. The paces are indicated by the previous caution of the commanding officer; but if no number is specified, six paces is the regulated distance between the files. If the left hand file, who leads on this occasion, be a steady man, and has moved correctly on his front, the line will require little dressing, too much attention to which, in extended order, is to be carefully avoided. If the distance between the files be not correct, it must not be altered by closing or opening out:—This can only be done upon the march.

To extend from the left.

5. Requires no additional explanation. The rear rank men cast their eye over the left shoulder.

To extend from the centre.

6. In extending from the centre, the left hand file of the right sub-division (if a company) is the centre file from which all movements take place. The other files face outwards, and proceed as before. When battalions extend from the centre, it will be performed quicker and more regularly by the companies moving in close order to the required distance, and then extending from the proper flank.

7. So

To fire in extended order on the spot. 7. So soon as the "*Fire*" has sounded, the rear rank men take a side step of ten inches to the right; and both ranks fire alternately in this position, commencing with the front rank; each making ready when he hears the ramrod of the rank which has fired, working.

To cease firing. 8. In this manner the firing is continued, until the bugle sounds the "*Cease*." After this sound, not a shot must be heard; the unloaded men re-load as rapidly as possible, and if any rear rank men should happen to be in front, they fall into their natural places, covering exactly as before; the whole then remain steady, and ready to move.

To fire kneeling on the spot. 9. This caution is repeated by the officers; and where there is any deficiency of them, by the serjeants, along the line. The bugle sounds the "*Fire*," and the whole drop instantly on the knee. The right knee is on the ground, and the right leg to the rear. The rear rank man, in coming down, disengages to the right, but not more than is necessary, that he may not be too much exposed. The firing proceeds as before, with this difference, that the rear rank men retain their place, and continue disengaged, to avoid the awkward movement of covering and uncovering upon their knees.

To fire lying 10. As soon as the bugle sounds the *Lie down* and the "*Fire*," the whole drop on both knees, (the rear rank men disengaging), and throw themselves on their bellies;—the firing proceeds as before; the men load on

thei

their knees, or they may load sitting or lying, though the latter is an objectionable position, and very liable to accident. Riflemen may fire on their backs in favourable situations; in this position, the feet are crossed, the right foot passing through the sling of the rifle, and the piece supported by it; but this position is not suited to the musquet of light infantry companies of the line: it furnishes a steady aim with a rifle, but it can only be used in cloudy weather or with the sun at the men's backs.—If in a very exposed situation, the soldier attempts to load lying, he will, after priming, roll over on his back, and, placing the butt between his legs, the lock upwards, and the muzzle a little elevated, draw his ramrod, and go on with his loading without exposure, rolling over on his breast again when ready to fire.

To form in chain order, or order of double files.

11. The caution being passed to "*Form Chain Order,*" the left files face to the right, and close upon the right files, without further word of command. The whole line now stands extended in double files, ready to advance by alternate files, if required. Light troops are never to remain halted and exposed unnecessarily at Chain Order, neither is any firing allowed in this formation. It is to be used only preparatory to the advance by files.

To re-form—order of single files.

12. On this caution, the left files face to the left, and take up their former situation.

The line will advance.

13. At the last sound of the bugle to "*Advance,*" the whole step off in quick time, dressing by the centre.

14. When

To fire—advancing.

14. When extended in single files, and *no mode of skirmishing is specified*, it will be by alternate ranks.—When extended in double files, it will be by alternate files. But skirmishing by files, whenever circumstances permit, will prove the most efficient manner, and should always be preferred, if possible—from the increased confidence with which it naturally inspires the soldier, more particularly in advancing.

To fire—advancing by alternate ranks.

15. The *Advance* and *Fire* having sounded, the front rank men give their fire independently as before. The rear rank disengages, moves on the number of given paces, in double time—(12, if covering an advancing line, and 24 if skirmishing detached,) deliberately aims at the enemy, and fires when the front rank has approached, and so on alternately.—The ranks wait for no signal to advance at the same moment, but whenever the man who has fired, is reloaded and ready, he moves on, and looks at his file-leader, and the enemy, without attention to his right or left-hand neighbour. In the field, a skirmisher, in advancing, regulates his distance to the front by the cover and advantage that the ground presents, and, if these are particularly good and commanding, he fires two or three shots without moving, as long as he sees that he retards none of his comrades. In firing, advancing, by alternate ranks, the rear rank men take care to advance always by the right of the men in their front.

16. In

To fire—advancing by alternate files. (when extended in chain order.)

16. In firing, advancing by alternate files, the rear rank men of the right files, instantly disengage to the right, and fire independently, but cover their front rank men when they load. The left files then move out, covering exactly, while the right remain halted, to load, and the firing is continued as before.—The attached officers move along their respective lines, attended each by a Bugler, and are not confined to any particular spot.—The captain, or other commanding officer, is usually in the centre and rear of the whole.—If the line of skirmishers is already in march when the "*Fire*" is sounded, the whole make a momentary halt, the right files give their fire, and the left move out, as already detailed.

The line of skirmishers will retire.

17. If the company or battalion in extended order be directed to retire by ranks, the front rank men give their fire, and go to the right-about; the rear rank men disengaging to the right, to let them pass.—Having retired the regulated distance (24 paces) in double time, they halt, and when loaded, the rear rank men give their fire, and retire beyond them, passing by their left.

To retire by alternate files.

18. If the company or battalion in extended order be directed to retire by alternate files, the right files fire first, the rear rank men disengaging for this purpose to the right, and go to the right-about; then the left files, who retire beyond them as before. The files cover exactly in retiring as in advancing.

19. The men come to the *left about* upon halting,

halting, and proceed with their loading. They come to the left-about, because the arms being at the long trail, would otherwise not be clear of each other.—In going to the right-about, after firing, the men go smoothly round upon their heels, without bringing them square;—they preserve the position at which they stood in the *present*, and have already, in coming about, one pace to the rear; the piece is brought at the same time to the trail.—They go to the right about because the rear rank men (whether retiring by alternate files or ranks) having disengaged, as above explained, renders no deviation from the prescribed mode necessary.

20. So soon as the ramrods of the rear rank commence to work, the front files, covering their opponents, fire independently as opportunities offer, go to the right-about, and retire as before.

To fire kneeling, advancing. (from extended order in single files.)

21. The caution having passed along the line at the last sound of the bugle to "*Fire,*" the whole drop instantly upon the knee; the front rank men give their fire, and the rear rank men spring up, and advance, by the right of their file leaders, the regulated distance to the front.

To fire kneeling, advancing. (from extended order in double files.)

22. If from chain order, or order of *double* files,—the right files first give their fire, and the left files spring up, and advance as before.

23. On the sound to "*Fire,*" the whole line or lines, drop instantly on the knee, the right files (or the front rank men only, if retiring

tiring by *ranks)* giving their fire, and then rising and going to the right about;—the left files (or the rear rank men only, if as before,) following as previously described.

To fire kneeling, retiring.

24. If already in two lines, when the "*Retire*" and "*Fire*" be sounded, the line which may be *then* in front, will, of course, be the first to give its fire and to retire.

25. It is an invariable rule that skirmishers always load *before* they advance, and *after* they retire, unless expressly ordered to do so on the march. If the "*Cease fire*" sounds without the halt, then the men load on the march.

To close.

26. On the signal being sounded, the men trail arms, face to the point required, and close in quick time.—If the double quick be sounded, they take it up upon the march, and shoulder, and dress, as soon as they reach the part to form upon.

To extend while a division is advancing;—from the centre—from the right—or, from the left.

27. In all these cases, the files from which the extension takes place, move straight forward in quick time; the others make a half turn to the flank, to which they are ordered to extend.—As soon as each file has got its regular distance, it will turn to the front, and advance; rear rank men covering their front rank men, and keeping in line with the directing file.

To incline to the right. Incline to the left. Advance.

28. The skirmishers make a half turn to the flanks to which they are ordered to incline, and continue in the diagonal direction, until the "*Advance*" is sounded, when they will return

return to their original front, and move forward as before. If, when the skirmishers have made the half-turn, the bugle should sound the "*Incline*" a second time, the men's shoulders should be brought up, so as to complete the face, and march in file.

To fire and load upon the march in a single line.

29. At the signal to "*Fire*," the front rank man of each file fires, and instantly drops to the rear, by the left of his comrade, and loads, as quickly as he can, upon the *march*; and so soon as his ramrod, begins to work, the other man fires, and proceeds in the same manner, taking care that both men are never unloaded at the same time.—This rule is always to be attended to, but more particularly in this instance, when the skirmishers are advancing in a single line and firing without halting.—This movement applies more particularly to a rapid advance upon a retiring enemy.—When regular resistance is encountered, the formation of two lines, together with the utmost practicable regularity in the alternate advance of each, is to be observed.

To halt.

30. At this signal, the whole kneel down, and take advantage of any inequalities of the ground near them, continuing to fire until the signal for "*Cease Firing*" has sounded.

To retreat.

31. If the signal to "*Retreat*" should be sounded when the skirmishers are *not firing*, both ranks will retire together; rear rank leading; but if firing at the time (which presumes always contact with the enemy) they will retire, as above directed, in two ranks, the front rank

rank men moving off first, and loading after they have halted and fronted; unless the "*Cease Fire*" sounds without the halt; (Vide Nos. 17 and 25 of the section.)

To halt.

32. If the "*Halt*" should be sounded, the rank next the enemy will stand fast, (or face about, if not already fronting to the enemy) and the other rank will close up to it, and the whole continue firing, taking care that both ranks are never unloaded, as before.

To change front: to the right, on the right file.

Double March.

33. The right file faces to the right, kneeling, the others rise up and trail:—At the word "*Double March,*" they bring the left shoulders forward, and form on the right file:—The distance will be preserved from the halted flank.—Each file will move in the shortest line to its situation in the new position, and instantly kneel down.

Distinction between changing front and throwing a wing backward or forward.

34. But in throwing a wing backward or forward, the distance of files must be preserved from the inward flank, and they must look to the outward flank for dressing, and bring forward the shoulders gradually, conformably to its progress.

To change front to the right on the left file.

Double March.

35. The left hand file faces to the *right*, kneeling. The others rise up, trail, and face to the right about; step off at the double march, and when in line face about and kneel.

To change front to the rear of the centre file.

Double March.

36. The centre file faces to the right about, and kneels; all the others rise up, trail arms, and face inwards by sub-divisions, and counter-march by files in extended order; the right sub-division passing in rear of the centre file,

file, and the left sub-division in front of it. Each file will kneel the instant it arrives at its place in the new line; or this may be performed by the sub-divisions bringing forward their shoulders inwards, each file passing through the alternate intervals, and halting as before.

37. The last four movements are principally intended to be practised at the drill, and with small bodies, in order to render the men intelligent;—but on service they can rarely be required. Whenever circumstances render these direct changes necessary in extended formations, it is desirable to throw out a new line of skirmishers from the reserve supports, or some other body.

S. 129. *Formation of the Chain, when skirmishing in front of an advancing Column—(formed right in front).*

1. The *Call* being sounded, succeeded by the "*Skirmish,*" the light company (*aa*) wheels outward by subdivisions, receiving the word from its own officer *Right and Left Shoulders forward,* and when in the due diagonal direction, *Forward.* It proceeds in quick or double quick time, as may be ordered, to skirmishing distance, *viz.*, 300 paces beyond the head of the column; the right sub-division on the right flank—the left on the left.

Fig. 65.

EXTEND. 2. The EXTEND being sounded, each sub-division extends, and forms in front of the advancing

vancing column; a section of each (a c) overlapping the flanks.

CLOSE.

3. When the CLOSE sounds, the skirmishers close to their respective sub-divisions.

ASSEMBLY.

4. When the ASSEMBLY sounds, these re-form company again, and resume their place in battalion.

General rule to be observed when skirmishing at a distance from the column or line.

5. But if the company be directed to skirmish at a distance, detached from the timely support of the column or line, one general principle must be observed, namely, that never more than one half must be sent forward to skirmish at a time; the other half remain formed and ready to support.

6. If a company therefore be directed to make an attack, or by means of the above disposition to keep the enemy at a distance from their front, the commanding officer, having arrived in position, will signify whether the right or left sub-division is to advance —If the latter, the left sub-division advances directly to the front, and when at skirmishing distance, extends from the centre, overlapping he flanks of its support, which remains ready formed in close order, and follows as the skirmishers advance. The captain commanding remains with the reserve.

Fig. 68.

7. If a battalion (a a) be employed for the above purpose, the left sub-divisions of each company move briskly 50 or 60 paces forward (a b); the right sections of those sub-divisions halt with closed ranks; the left sections move the same number [or any directed number]

number] of paces (*c*) further to the front, and extend their files from the left, so as to cover completely the front of the main body from which they are detached; the outer sections of the battalion overlapping the flanks. Whenever right sub-divisions advance to skirmish, the right sections must be pushed on to their front, and extend themselves from the right, and *vice versâ*.

8. In this order of formation, and which would obtain equally with the single company, if strong enough for the purpose, the intermediate halted sections (*bb*) are called the supports, and the rear halted sub-divisions (*a a*) the reserves.

9. When a light battalion marches in open column in the vicinity of an enemy, one company will be formed in front, as an advanced guard; one in rear, as a rear guard, and half a company on each flank, as flanking parties.

10. The same disposition to extend to a brigade of the line, when the three light companies may be employed in the same manner.

Flanking parties.

11. The company to perform this duty will extend its sub-divisions, so as to cover the entire flank or flanks of the column.—If intermediate supports are necessary, which can alone be determined by circumstances, and the relative distance at which the flanking parties act from the column, they must be furnished [in the absence of light troops] *from the line*. These supports will be formed about 100 yards in rear of the skirmishers, and

and the whole will move in file parallel to the column, and the leading file will take the outside flank skirmishers of the advanced guard for his general line of march. When the column halts, the flankers and supports face outwards to the enemy.

How the unity of companies, when skirmishing, is best preserved

12. With a view to preserve as much as possible, the unity of companies when employed in skirmishing in an open country, or in the presence of cavalry, it is always desirable when circumstances permit, that the skirmishers and the supports should be composed of the same companies, and the reserve formed of distinct companies. By this means, each company is more easily united, and throws itself, if attacked, more readily into square.

13. When this arrangement is adopted with a battalion, it is only necessary to fix the proportion for the reserve, which, if circumstances permit, should never be less than one-third of the whole body. The right subdivisions of the remaining companies are then ordered forward to skirmish, while the left sub-divisions form the intermediate supports The discretion and intelligence of commanding officers will readily apply both the modes of formation laid down. The principle throughout is the same, but as each company may have to act singly, it is necessary that it should be prepared by the practice of the first method to form its own reserve, and to depend exclusively upon itself, whenever circumstances require

S. 130. *Advanced Guard.*

On the line of march.

1. When a column is marching along a road where it is not considered necessary for the advanced guard to form a chain of skirmishers, the company will be formed in sections; the two rear sections [under the command of the captain] will form the reserve, advanced in front of the column; the second section from the front [under the command of a subaltern] will be 200 yards in front of the reserve, and the leading section will be 100 yards in front of the second section, and will detach a double file, under the command of a corporal, 100 yards in its front, and a double file to each flank, 100 yards from the road, and about 50 yards more retired than the corporal's party. These detached files must carefully examine all houses and enclosures within their reach; but should more distant objects present themselves, patroles must be detached from the second section for their particular examination. Single files of communication will be placed between the different divisions, and also between its reserve and the head of the column. The distance between these two last must be regulated by circumstances; but it will be commonly estimated at about 500 yards during the day, and about 300 during the night.

Patroles.

2. Patroles may consist of a subaltern's party, or of a serjeant and twelve, or a cor-

 poral

poral and six men, according to circumstances.

3. The object of a patrole is to obtain intelligence, and to ascertain the presence or position of the enemy.

4. It is a general rule, therefore, that a patrole never commits itself in action if it can be avoided, but retires [under cover, if possible] as soon as the requisite information is obtained. On coming to a house, an enclosure, or a hill, a single file of the patrole will advance and examine it; another file remaining behind, will watch its motions, and be ready to give assistance, supported by the reserve, if required. As soon as the file in advance is satisfied that there is no enemy in the place, one of the men will make a signal, by holding up his firelock above his head, in a horizontal position, and the rear file will join and move forward as before.

5. On approaching a village, the same precautions will be observed in front, while flanking parties will move round the outskirts.

6. In passing a defilé or hollow way, (in order to guard against surprise), a number of files of the patroles should follow each other in extended order, each file keeping the preceding one in view; and flankers must, if possible, examine the ground on the right and left for further security.

Advanced guard in a plain in presence of the enemy.

7. When a detachment of light infantry is ordered to form a regular advanced guard in

in a plain, in presence of the enemy, it will be divided into four sub-divisions, or sections, each of which must be at least 16 files. Each section, moreover, will require at least an officer, a serjeant, and a corporal, independent of two centre serjeants to lead the whole.

Fig. 66. 8. The detachment (*dd*) being in line with the centre serjeants in the centre, and the officers on the right of their respective sections, receives the order to "*form the advanced guard.*" The officer commanding the third section immediately gives the word, "*Trail Arms,*"—"*Quick or double March,*" and leads out 100 paces directly to his front, upon points he shall have previously selected. The officers commanding the first and fourth sections move on at the same moment, giving the word "*Right and Left Shoulders forward,*" and march the same distance, in a diagonal direction, upon points furnished by themselves, in the same manner. The second section remains in reserve with the officer commanding the whole. The centre serjeants move to the front with the third section, as before stated. As soon as the three sections are halted, they establish their chains of communication to the flanks and rear, in the following simple manner:—

No. 1 section, throws out one file 30 paces on its own right, and another file 30 paces on its own left (*aa*).

Nos. 3 and 4, do the same.

In addition to this, No. 3 throws one file (*b*) 30 paces to the rear, to keep up the

 communication

Fig. 66. communication with No. 2 in reserve; and No. 2 throws forward one file (*c*) the same number of paces, to meet the file of No. 3.

Having done this as rapidly as possible, the outward half sections of Nos. 1, 3, and 4, advance exactly as before, (on a signal from the centre), but only half the number of paces, and the moment they halt, extend in skirmishing order.

No. 1 from the right,

No. 4 from the left,

No. 3 from the centre,

which continues to be marked by the centre serjeants. An officer is posted with the line of skirmishers; the others remain with the supports, and the captain with the reserve of the whole. As soon as the skirmishers extend, (which they do at the distance of 10 paces), the supports in their rear open a communication with them, by pushing forward a double file (4 men) 25 paces in their own front (*e e e*).

9. The advanced guard is now formed; the officer (*o*) commanding the line of skirmishers takes post between the centre serjeants, and is ready to lead the whole; although when required, his presence is confined to no spot in particular. In advancing, the whole dress by the centre serjeants. The double files (*e*) of communication from No. 3 *Support*, cover them exactly, and by this means furnish a point of conformation for he rear.

10. Should circumstances render the presence

sence of the centre serjeants necessary to command the supports; No. 3 throws forward its right file (intelligent men) as the file of direction. If the advance guard has a path in its front, leading to its destination, this file of direction should be placed on it, and the men told to move by it; the other files conforming to the directing one. When the road takes a bend of any consequence, the directing file will mark time whilst the shoulders of one flank are brought forward, and the other flank halts. When this change of direction is completed, the leading file moves on again.

S. 131. *To pass a Bridge or Defile.*

Advancing. 1. If the advanced guard has to pass a bridge or defile in front, the skirmishers gradually draw inwards as they approach it, then run forward and close up their files, followed by the supports and reserve; and as soon as they have passed it, they will extend in their former order from the right, centre, or left, as circumstances require.

Retreating. Fig. 67. 2. To pass a bridge or defilé retreating, the reserve and supports first pass through, and the former extends as a new line of skirmishers, (*ee*) while the supports (*s*) in close order form at the end of the defilé, ready to fire upon the enemy, and protect the skirmishers until they pass; the latter (*cc*) draw inwards by degrees on their retreat until they get

get close to the defile, when they run through, pass 50 or 60 paces to the rear of the new skirmishers, and form as a reserve for *them*(cc).

Fig. 67.

How to pass a wood in extended order.

3. In passing through a wood, the skirmishers will draw inwards, or open out, according to its extent; the outer flank files occupying the skirts of it, but keeping sufficiently within the wood, as not to be perceived from the outside.

How to break through a copse, or wood, when in line.

4. Should a corps of light infantry, *when in line*, have to break through a thick copse or wood, where it is impossible to preserve any order, every individual must follow the openings which may appear best: in such situations, however, the officers and serjeants must be the first who get through it, that each company may form upon them as quickly as possible.

5. Should the advanced guard meet with bogs or other impassable ground of any extent, they must not leave it between themselves and the column, without careful examination, for fear of being cut off, or an enemy being concealed within the chain.

Rear guard.

6. A rear guard is an advanced guard reversed; and the principles and instructions given for the formation of the latter apply equally to it.

S. 132. *Picquets and their Sentries.*

1. The strength of picquets will depend upon the ground they are to occupy, their distance from, or proximity to, the enemy, and the importance which may be attached to the posts they are intended to defend.

Principal duty of a picquet.

2. The principal duty of a picquet is to guard the army in its rear from surprise, and to oppose such small detachments as the enemy may push forward for the purpose of reconnoitring, &c.

How posted.

3. To accomplish these objects, the picquet must be posted contiguous to the principal road it has to defend, and if possible, behind some cover, to conceal it from the view of the enemy, taking care, at the same time, to leave no road unobserved on its flanks, by which it might be surprised in its rear.

Measures for immediate security.

4. As soon as the picquet arrives on the ground it is to occupy, sentries will be thrown out in elevated situations upon its flanks and front, for *immediate* security, and the officer commanding, having reconnoitred the neighbourhood of his post, with a strong patrole, in person, will proceed to form a chain, covering his front and flanks, and communicating with the picquets on his right and left.

Line of sentries.

5. In selecting the line for the chain of sentries, care must be taken not to extend it too much,—to post the men in the most advantageous situations for observing the roads and country in front, and to keep them

as

as much concealed from the view of the enemy as the nature of their duty will admit. Sentries must be so placed, moreover, as to secure one another from being cut off, and at such distances as to prevent any enemy from passing unperceived between them during the night.

Sentries of advanced picquets to be invariably double.

6. To ensure this object, and to guard against surprise of every kind, the sentries of an advanced piquet will be invariably posted double, and one of them will be always walking to the right, till he approaches the adjoining file, while the other is looking vigilantly out to his front. In this mode they alternately relieve each other.

Detached parties under non-commissioned officers.

7. If the chain of sentries should be so far extended as to make it inconvenient to relieve them all from the main body, one or two small parties, under the command of a non-commissioned officer, may be detached to a convenient situation, for the purpose of furnishing the sentries, and forming an immediate support to the chain.

Advanced picquets require three reliefs.

8. All advanced picquets should have three reliefs.

When firing is heard.

9. When any thing particular is observed in front, or firing heard in any part of the line of picquets, either by day or night, one of the sentries will instantly run in, and report to the officer what he has noticed, taking care to mention the circumstance at any post he may happen to pass by on his way.

10. When

When piequets are attacked.

10. When piequets are attacked, the same rule will be observed as in all other skirmishing, and the detached serjeants' parties will not run in on the main body, but support the skirmishers; and when compelled to retire, they will, if possible, retreat on the flank of the main body, and thereby afford mutual support to each other.

When posted in a village

11. If a picquet should be posted in a village, the main body must be placed, so as to be behind the junction of all the roads that lead to the enemy's position, the entrances from which must be blocked up, or dug across, with the exception of a small retiring path for the sentries placed in observation in front. Small parties will be placed behind each barricade for its defence.

First duty of an officer on picquet.

12. The first care of an officer ought to be the strengthening of his post, by constructing abbatis, breast-works, &c., and particularly where the defence of a bridge or ford is intrusted to him, he ought never to omit to throw up something of the kind to protect his men, and impede the advance of the enemy.

A picquet must not shut itself up without orders.

13. A picquet ought not to shut itself up in a house, or an enclosure, with the intention of defending itself to the last extremity, unless particularly ordered to do so, or that circumstances may render it necessary at the moment, for the preservation of the party, in the expectation of support.

14. A

Under what circumstances a picquet should retire.

14. A picquet may with safety defend its front as long as its flanks are not attacked, but as soon as the enemy attempts to surround the post, the picquet must begin to retire.

Signals by day.

15. Signals may be established by sentries during the day:—For instance—one man holding up his cap on the muzzle of his firelock, signifies that the enemy's patrole is advancing; and both men holding up their caps in the same manner, signifies that the enemy is advancing in force.

Flags of truce.

16. On the approach of a flag of truce, one sentry will advance and halt it at such distance as will prevent any of the party who compose it from overlooking the picquet-posts. The other sentry will acquaint the officer commanding the picquet of the circumstance, who will, according to his instructions, either detain the flag of truce at the out-post, until he has reported to the officer of the day, or he will forward the party blind-folded to the camp, under an escort. If the flag of truce is merely the bearer of a letter or parcel, the picquet officer must receive it, and instantly forward it to head-quarters. After having given a receipt, the flag of truce will be required forthwith to depart, and none of the picquet must be suffered to hold any conversation with this party.

Night duties.

17. At night the situations of the sentries ought to be changed, and generally drawn back

back nearer the supports, and placed so as to have the high ground before them; as an object is more easily discerned at night from a low situation, than when looking from a hill.

18. The sentries ought, if possible, to be relieved every hour during the night.

The countersign only given to the sentries.

19. When a sentry is posted, the countersign only is given him; and no person under the rank of a non-commissioned officer is usually intrusted with the parole, which serves as a test for passing armed bodies inside the line.

Challenging a double sentry.

20. The moment a sentry sees or hears any person, he calls out, "Halt,"—"Who comes there?" and at the same time ports his arms, fronting to the party. If a double sentry; one stands behind the other, uncovering to the right, and also comes to the port. If the answer be not satisfactory, the leading sentry instantly fires.

Deserters.

21. Sentries must be very distrustful of people who answer the challenge by saying "Deserters;" they must be immediately ordered to lay down their arms in the rear, and not suffered to approach the sentry, until a party arrives from the support to receive them, and then singly. If the suspected person hesitate, the sentry will instantly fire.

Advanced picquets to be under arms an hour before day-light.

22. Piquets will get under arms in the morning an hour before day-light; and if every thing appears quiet in front, the officer will, as soon as he can discern objects distinctly, proceed to occupy the same posts that he held

held the day before; but he must previously send forward patroles to feel his way, and should any change be remarked in the enemy's posts or position, he will report it immediately to the officer of the day.

When advanced picquets should be relieved.

23. As attacks are most commonly made about day-break, a desirable accession of force will be always obtained by relieving the piquets at that hour.

Arrival of the relief.

24. When the new picquet has arrived, the officer commanding it will accompany the officer of the old picquet along the chain of posts, and this officer will point out the situation and strength of all the enemy's posts, and afford every other information to the relieving officer, in his power.

Duty of the offiecr of the old picquet.

25. When the sentries are relieved, and the weather is sufficiently clear to ascertain that there is no indication of an attack, the officer who has been relieved, will forward a written report to the officer of the day, and march his picquet home, *but if the advanced picquets* should be attacked before he arrives in camp, he will consider it his duty to face instantly about, and march to their support.

S. 133. *On the Sound of the "Assembly."—How to run in upon the Battalion, according to its Situation,—and resume their place in Line.*

1. On such occasions it is of the utmost consequence that the front of the battalion should be left clear as soon as possible.

2. The

2. The skirmishers, therefore, if detached to any distance, must endeavour instantly to discover the exact situation of the battalion, and decide in what direction to run in, adopting that mode which will least impede, and soonest leave it in a situation for firing or advancing.

If marching in line.

3. If the battalion is advancing or retiring in line, they will run towards each flank of the battalion, separating from the centre.

If in echellon.

4. If in echellon, they will proceed towards the outer or reverse flank.

If throwing a wing forward or backward.

5. If throwing a wing forward, or backward, they will make for the outward flank.—In both movements, the inward flank (or that which is first formed) will be left clear for firing, and in the first movement, they will have less distance to run over.

Open column to close column.

6. If the open column is forming in close column, they will run in towards the rear division.

Close column to open column.

7. If the close column is forming open column, they will run towards the reverse flank, proceeding round the standing division of the column, or passing through the divisions as they open out, if necessary.

Line counter-marching.

8. If the battalion in line is counter-marching, they will open out from the centre, and run round the flanks, forming in rear of each, and afterwards closing.

Formation of square.

9. If the battalion forms square, they will take the most direct and short way to the rear, and close up and compose the rear face.

S. 134.

S. 134. *Changes of Skirmishers, when not called in, to correspond with the Movements of the Battalion.*

1. If the skirmishers *are not called in* while the battalion performs any movement, they must, with the utmost rapidity, change their situation, so as to correspond with the new order of the battalion; and their attention and activity are chiefly required in protecting it during the change.

Line breaking into open column.

2. If the line breaks into open column, they must face to the right, or left, and take ground to the same flank.

Column counter-marching.

3. If the column countermarches, they must face to the right about, and move rapidly along the reverse flank to the new front.

If the column wheels into line.

4. If the column wheels into line, the skirmishers must also change their direction, and extend along the front of the line.

Open column to half or quarter distance.

5. If the column closes to half or quarter distance, the skirmishers must also decrease their distances between their files, and some of each Section (when skirmishing at a distance from the line) must be called in to the supports.

Column changing front.

6. If the column changes its front by the successive march of divisions from the rear, (which will change the front of the line when wheeled up) or if, from any other cause, it becomes necessary to shift the skirmishers from one flank of the column to the other, they will run through the divisions, and pass to

to the other flank, preserving, as nearly as they can, the order of their files from right to left, which should always be retained, or renewed as soon as possible.

Battalion forming square.

7. When the battalion forms square, and the skirmishers are not called on to make a component part, they will form the rallying square, in echellon with the square or squares of the battalion.

8. If suddenly attacked by cavalry, the skirmishers will take shelter in the square, forming a reserve, which will be ready to move out or fill up vacancies as required.

S. 135. *Rallying Square.*

1. When a company at extended order, and skirmishing detached, is suddenly surprised by cavalry, the "*Alarm*" will sound, followed immediately by the "*Assembly*."—This will be considered as the signal to form the rallying square as laid down for the line in general. *Vide S.* 67.

Fig. E.

2. When the reserves (Vide S. 129, No. 12,) are attacked by cavalry, they may each form three deep, and wheel into a square composed of the four sections; or they may form grand division squares as instructed in S. 123.

3. When the supports are so attacked, they may each form an orb by forming three deep, and wheeling backwards into a circle; and the skirmishers as they arrive will form round them, unless they may have previously formed a rallying square.

BUGLE

BUGLE SOUNDS.

No. I.—*To Extend.*

No. II.—*To Close.*

No. III.—*To March; also Forwards, after obliquing.*

No. IV.—*Halt; also to Annul; annuls every previous sound excepting* No. V.

No. V.—*To Fire.*

No. VI.—*To Cease Firing.*

No. VII.—*To Retreat.*

No. VIII.—*To Assemble.*

No. IX.—*To Disperse.*

No. X.—*To Skirmish.*

No. XI.—*Incline to the Right.*

No. XII.—*Incline to the Left.*

No. XIII.—*The Alarm.*

No. XIV.—*The Lie Down*

No. XV.—*The Rise.*

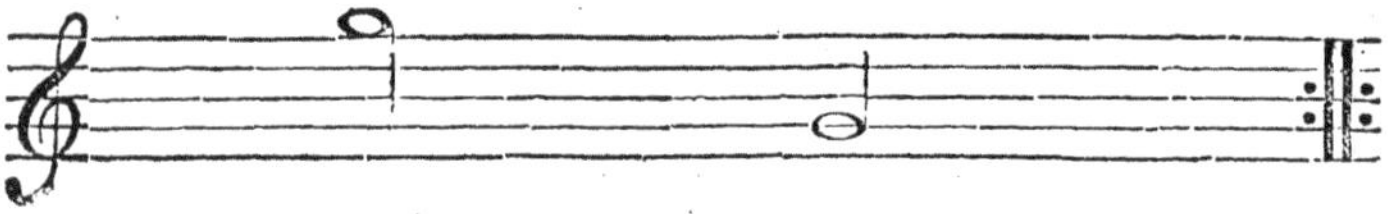

The Light Infantry Call.

The Assembly of Officers.

The Quick Time.

The Double Time

End of Part Fourth.

PART V.

GENERAL PRINCIPLES

FOR THE

MOVEMENTS OF THE BRIGADE OR LINE.

(Plates IX, X.)

A battalion may be considered to hold in a line the relative situation of a company in a battalion; and it will be found, that the chief part of the General Principles laid down in Part III, for the movements of the battalion, has necessarily embraced a more extended application to the movements of a brigade or line: It must thus follow, that a careful study of the sections of which the Battalion General Principles are composed, will qualify an officer for their application to more extended bodies; and therefore the inconvenience of repetition may be avoided in the following sections, where such general principles only will be treated on, as have not been provided for in the Battalion Principles, and which exclusively appertain to the movements of greater bodies.

S. 136. *Formation and Distances.*

1. Great bodies of troops are formed into one or more lines; each line is divided into right and left wings; each wing is divided into divisions; each division

division into brigades; and each brigade into two, three, or four battalions.

Distances in time.

2. Battalions should be formed in line at the distance of six paces from each other: No increased distance need be allowed between brigades, unless for the admission of guns, which may be posted on the right of brigades or divisions, according to the disposition of the commander; but never between the battalions of a brigade, unless particularly ordered.

3. Previous to treating upon the distances of column, it may be necessary to state, that contiguous columns are called at close order, when they are immediately contiguous with small intervals between battalions: They are called at open order, when at the distance from each other at which they are formed from line.

Distances in open Column.

4. When divisions or brigades are formed in open column of route or manœuvre, the distances between battalions should be equal to the front of the column and the six paces which separate battalions when they wheel up into line: And the leading company of each battalion should therefore preserve the latter distance, in addition to that of its own front.

Distances between Battalions in Column at quarter distance.

5. When a column of route or manœuvre is formed in mass at quarter distance, the distances between battalions should be ten full paces, in order to leave room for the pivot file of the leading division of each battalion to advance six paces, to facilitate the wheel into contiguous columns: (Vide S. 139, No. 2). But when contiguous columns at quarter distance are formed at close order, there

is

is no occasion for a greater distance between battalions than six paces. (Vide S. 139, No. 3.)

In close Column. 6. The distances between battalions when formed in mass, or in contiguous close columns at close order, will always be six paces: No increased distance need be allowed between divisions or brigades formed in mass, or in contiguous columns at close order, except for the admission of guns: But it must always be recollected, that the distance of ten and six paces will invariably be required, in addition to the space which the artillery may occupy.

Figs 86, 89 7. All distances between battalions in column will be taken up by adjutants, (*a*) who will mark where the proper pivot flank of the leading division of the column is to rest. All points of entry for columns to march upon these adjutants, will be taken up by field-officers. (Fig. 89.) In formation of lines, by deployment and echellon the adjutants of battalions are required to mark the outer flank of each, in order to connect battalion with battalion in their true prolongation, (Vide S. 74, No. 2,); therefore the distances between battalions in line should be marked by field-officers from the adjutant of the next battalion. In order that mounted officers may give their essential assistance to battalion commanders, they should always be ready to arrange these points with activity and intelligence.

S. 137. *Commands.*

1. It is of the first importance that all words of command to a division or brigade, should be circulated with the utmost precision and rapidity, for

which essential purpose the eye and attention of each battalion leader must be constantly directed to the commanding general, or to the regulating battalion, when he cannot be heard. The general commanding must give his short orders of caution to the commander of the regulating battalion, or to whatever battalion is nearest to himself, in such terms as may not be mistaken by the soldiers for a battalion order. The battalion commander will proceed without delay to execute the order in the usual terms of command, and in so loud a tone of voice as to be heard by the contiguous battalions.

2. The execution of a command may be regulated by a signal, such as a trumpet, or any other method for effecting simultaneous movement, which the commanding general, at his discretion, may fix upon.

3. Where the wind, or the noise of fire-arms, may interrupt the progress of an order from battalion to battalion, the commanders will conform as quickly as possible to the movement which they see executed to their right or left; and staff officers will proceed promptly to convey the general's orders to the corps which, from these causes, may not have heard them pass from the regulating battalion: For this essential purpose staff-officers may be attached, to carry orders to particular bodies, so as to prevent the possibility of confusion; and where they are not in sufficient numbers, the different field-officers of regiments, who may be placed at intermediate stations, will assist in passing the order from one battalion commander to another: But great care must be taken by those who pass an order, that they have properly understood

it

it themselves, for much mischief and confusion might arise from a multifarious communication of orders imperfectly delivered; and to prevent this the more effectually, it may be, in certain cases, expedient to send the command by a staff-officer, in writing.

4. When troops in extent of numbers are at all beyond the immediate control of the voice, it is particularly essential that the commands to HALT, or to MARCH, should be communicated by a signal, upon which the battalion commanders may act simultaneously; for the accuracy of combined movement would be totally lost, if the different bodies in line, or in column, be suffered to march, or to halt, as their respective commanders may catch the word in its approach from a distant quarter.

S. 138. *The March in Line.*

1. The march of a line to the front, when it consists of several battalions, is conducted upon the same principles which are pointed out for the guide of a battalion. The directing serjeants will, upon the caution, move out in the same manner; and one battalion will be named to regulate the direction of all the others.

Directing Serjeants.

2. The directing serjeants of the regulating battalion will be considered the guide by which the perpendicular direction of the whole line will be conducted. Their squareness of front being once established, as pointed out in S. 70, they must be

be preserved in it, under the careful superintendence of the commander of the battalion, who will have received, from the general commanding, his instructions as to the direction of the line.

3. The centre serjeants of the other battalions will conform to those of the one which regulates the march, and they will therefore be liable to change their perpendiculars of march, whenever they shall appear to the commander of the battalion to which they belong, to depart from the true parallel lines of direction with the regulating battalion, whose march is supposed to be accurate.

4. It is indispensable that the march of the regulating battalion should be conducted with the utmost exactness; it is the helm which guides the line, and it must preserve an unvaried cadence and length of step, which nothing but unavoidable necessity should interrupt.

5. It is essential towards the correct march of a line, that the injunction contained in S. 137, No. 4, should be observed, regarding the command of MARCH and HALT, which, in such movements particularly, should be executed by all the battalions of a line at the same instant, or else inaccuracy, and perhaps disorder, must ensue.

6. The whole attention of the commander of the regulating battalion must be given to the square and correct march of his battalion, in the direction pointed out to him: All other commanders must conform to his march; and if their centre serjeants swerve from the line, they will be prepared to order them to bring forward their right or left shoulders, as occasions may call for corrections, to preserve the

the true parallel lines, with the perpendiculars of march, on which the regulating battalion is moving.

S. 139. *Wheeling of Columns.*

1. The wheel of columns, according to the principle explained in S. 77, No. 5 and 6, is essentially useful, when applied to the different battalions composing a division or brigade, in order to effect a rapid change of formation or position.

Wheel of a mass of Columns into a Line of contiguous Columns. Fig. 69—A.

2. A mass of columns may always wheel into a line of contiguous columns, giving them the respective distances between battalions of ten and six paces, according as the columns may be at quarter or close distance, (Vide S. 136, No. 5 and 6): This can be effected without difficulty, because in the wheel, which should always be made simultaneously, each battalion removes its depth (*d b*) from the one which is wheeling in its rear, (*b*), still preserving the relative distances between battalions (*c c*), provided they are of equal strength.

Wheel of a Line of contiguous Columns into a mass of Columns. Fig. 69—B.

3. A line of contiguous columns at close order, and at quarter distance, can *never* be wheeled by battalions into a mass of columns, because the depth (*e e*) of each, must wheel round upon the pivot of the battalion on the right or left: And this wheel is equally impracticable for battalion close columns, unless the breadth of each column (that is, its front,) is equal in extent to its depth.

Fig. 69—C.

4. The depth (*a b*) of battalion columns must always be the same, in proportion as the establishment is of eight or ten companies: Thus, allowing

Fig. 6—9 C. ing one pace between ranks, and two paces between companies, a battalion of eight companies must occupy a depth of 22 paces, each of 30 inches, from the heels of the front rank men of the leading company, to the heels of the rear rank men of the rear company: And a battalion must be of sufficient strength to equalize the companies to 31 or 32 file, in order to give a breadth, or front, equal to the depth of 22 paces. Thirty-one file, each occupying 21 inches, will give 9 inches less than the depth; and thirty-two file will give 12 inches more than the depth. It may not be very material which is laid down as a datum; but it must be recollected, as a general principle, that the front of battalion close columns should be 31 or 32 file, in order to wheel from contiguous columns into mass.

5. If the column were wheeled in its square form, the space required for the movement would be the circle described by the diagonal (*b c*); and in that case the outward angle would come into contact, more or less, with the pivot of the adjoining battalion: But the column in its wheel, as laid down *Fig.* 49. *Pl. VI.* in (S. 109, No. 1,) becomes elongated in the one diagonal (*e o*), and compressed proportionally in the other (*a d*); by which it will not only wheel clear of the adjoining battalion, but be made to complete its movement, without spreading at all in depth.

6. When it is necessary therefore to wheel contiguous columns at quarter distance into mass, (or close columns, under 31 file for their front,) it must always be done when they are at open order, and they will then close upon the leading column to the

the distances required; or else, if formed at close order, they must open to a distance sufficient to admit the wheel.

7. It has been already observed, in (S. 77, No. 5,) that the wheel of columns is essentially useful for the rapid changes of front of a line of contiguous columns to the rear; and this will be laid down in more detail in the Evolutions: But it may be further observed, that the wheel is also applicable to the changes of front to a flank of a line
Fig. 69—D. of contiguous columns at close order; recollecting, that when the front is to be changed to the left flank, the battalion columns (*a c*) must move in echellon from the right (*a*), each battalion following (*b b b*) when the one in front has gained the extent of its breadth and the six paces of interval between battalions. When the front is to be changed to the right flank, the echellon movement must be made
Fig. 69—E, from the left: They should be halted when at the point where line is to be formed, and wheeled inwards by battalions (c *e*), which changes the front of the echellon, placing the former rear battalion (*c*) in advance, and the line of columns will then be formed upon it.

Base Points. 8. In all formation of lines of contiguous columns upon any fixed battalion, by the operation of the wheel, countermarch, or otherwise, base points must be established in front of the flanks of the leading division (c) of the named battalion, which will be taken up successively by the other battalions, according to the detail pointed out in S. 74, No. 7. It must be borne in mind, as a general principle, that whenever any body is named as an appui of a formation, it will always throw out these base poiuts withont without waiting for specific instructions to that effect.

S. 140. *Column of Route.*

1. The instructions laid down in S. 76, from paragraphs 31 to 36, for the march of a column, apply in an equal degree to considerable columns of route consisting of several battalions; for the greater the force, the more essential it becomes to preserve the regularity of movement, which is equally conducive to military success, as to the protection of the soldier from the fatigue of harassing efforts to regain lost distances, &c.: A compact and well regulated order of march, without straggling, forms a great criterion of the discipline of an army, and of the efficiency of its officers.

2. In all route marching, the baggage should be invariably formed on the reverse flank: No impediments whatever should be permitted in the line of march, except the cannon, which will generally be formed between the divisions or brigades o an army. When the admission of a portion or the baggage of an army into the line of march becomes unavoidable, it should be placed between the brigades, but never between battalions of a brigade.

Original front to be preserved.

3. The preservation of the original front of a column is of great importance: Unnecessary doublings should therefore be avoided; and when absolutely required, the original front of march should be resumed the moment the impediment has been passed.

4. A

Fig. 79

4. A body of troops which has a considerable march to make, will generally approach an enemy in one or more columns, at open or other distances, according to circumstances: Some general knowledge of an enemy's situation must determine the manner in which he is to be approached, the composition of the columns, the flank of each which leads, and their combination in forming: A nearer view may determine a perseverance in the first direction, or a change in the leading flanks and direction of the columns, in order to form in the speediest and most advantageous manner.

Importance of exactness in the march.

5. The success of military operations frequently depends upon the well combined march of columns, upon their several points of direction; and unless the greatest exactness is preserved in a column of route, no calculation can be made upon the time when a force, consisting of several detached columns, may concentrate at a given point, for a particular purpose: This exact movement of columns to the point of concentration is of the first importance in all military enterprises, particularly in all great movements of advance to attack, or to gain a particular position; and the utmost attention is required from all the component parts of an army, to support the regularity, which is essential to effect this combination of movement.

General objects in marches to the front.

6. During a march to the front, some one column should determine the relative situation of the others, upon the principle of the regulating column in manœuvre: But the separation of columns may unavoidably be considerable, according to the roads, or the means which the face of the country affords for their respective advances: As they approach

Fig. 89.

the

Fig. 79. the enemy, however, they must be so regulated, and directed, as to occupy the intermediate spaces when required to form.

General objects in marches to the rear.

7. The safety of marches to the rear must depend on particular dispositions, and on the judicious choice of such posts as will check the pursuit of the enemy. In these marches to front and rear, the divisions of a second line generally follow or lead those of the first, and all their formations are relative to them; but the disposition on these occasions must depend on local and temporary circumstances, which govern the discretion of the commander. Fig. 77.

S. 141. *Echellon of the Line.*

1. Echellon formations and movements are conducted in a considerable line upon the same principles laid down in Part III: they are calculated to place a great corps in an advantageous position to gain an enemy's flank; and sometimes they are formed with effect from the centre of a line, by refusing each wing (*b b*). If an attack made by an advanced corps (*a*) of a great echellon be effectual, each succeeding one moves up to improve the advantage: But if it fails, the succeeding bodies (*c c*) are in a situation to protect the retreat; and in gradually retiring upon each other, they afford a mutual aid and support. Fig. 70, 73. —74.

Direct echellons.

2. Direct echellons are formed by the successive advance of brigades, battalions, or half battalions, from either flanks, or from the centre of a line; or by columns placed in readiness for deployment in echellon Fig. 70.

echellon parallel to the enemy's position; and in both cases, the distances of at least half their front, will be preserved between the echellons, in order to give sufficient room for the mutual protection of flanking squares (Vide S. 142, No. 26), when such formation is required.

Oblique echellons.

3. Oblique echellons of a considerable line are formed for the purpose of gaining ground to a flank; and as it might be hazardous to throw the whole line into an echellon of companies, echellons of brigades, battalions, or half battalions, may be formed previous to a general advance, by the wheel and formation of companies on their respective flanks
Fig. 71. (*a a a*), as directed in S. 117. When the line has gained the ground required to a flank, it can resume the former parallel front, by the echellon march of companies in the same manner, upon the other flanks (*b b b*) of each brigade (b), battalion, or half battalion: In the formation of these great echellons, and upon wheeling up again to form a general line, each flank company (at *b*) should accurately wheel forward the same number of paces; and upon re-forming line, they should be correctly dressed upon the directing flank, before the other companies are moved up to form upon them.

Fig. 72. 4. When echellons are formed from an oblique line (*o b*), so as to stand parallel to an enemy's front (*b c*), the flanks of each echellon (*d d*) will be doubled behind each other, in proportion to the wheel made by the flank company (b) of each battalion: If this defect in the formation has not been previously provided for, by taking sufficient intervals between battalions when forming the oblique line, it must be rectified by taking ground to the flanks when in echellon, in order to place each body in the situation it would have held if formed from a parallel line. If this doubling of the flanks be not previously avoided, or remedied, a general line cannot be formed by the direct march of the rear echellons without excluding a part of each: But the line may be formed upon the leading echellon, by the rear ones moving (*c a*) on their respective adjutants, (*a a*) into the general line (*a* b), by the echellon march of companies to the reverse flank.

Fig. 73. 5. When oblique echellons of battalions (*a g*) or half battalions are to form a general line (*a* B) in advance, upon the leading division (*a*) of the front battalion, each of the rear battalions must wheel (*g h*) what will be in addition to their original wheel (*e g*) into echellon, one half of the total wheel *e g*,*g d*) made to the front (d), from line (*aa*), by the flank division of the leading battalion. Thus the leading divisions of the rear battalions will be placed upon perpendicular lines of march to their respective points of appui in the intended line (*a* B); the rear battalions will then advance in line until their leading flanks (*c c*) shall reach the alignement, and they will then halt and form up by the echellon march of divisions as in S. 117

Fig. 73.

Fig. 70, 73. 6. When the object is to gain an enemy's flank (*e*) the whole line is thrown into echellon (either direct or oblique, as may suit the case) from the flank (*o*) opposed to that of the enemy which it is intended to attack or turn; care being taken that the outer flanks of the echellon are protected in such advances from the enfilade of the enemy. When it it is necessary to refuse a flank attacked by an enemy, the line will be partially thrown into echellon direct to the rear, to cover one's own flank: But Fig. 74. when the enemy's attack (*a*) upon such flank (*b*) is repulsed, a counter attack may be made upon him to advantage, by an echellon movement from the other flank (*c*).

7. When the flank of a line is refused by the partial retreat of echellon battalions in this manner, the retreat, to the extent it is intended to refuse the line, should be made (if possible) upon a Fig. 70, 74. strongly posted flank (*d*), from which the fire of artillery could enfilade the advancing enemy.

8. In an open country, echellon movements to attack or to gain a flank are attended with difficulty and hazard, when the enemy is in a position from whence he can observe the intention: Advantage must therefore be taken of local circumstances that afford the means of partial concealment, or points of appui for the attacking body: All echellon movements against an enemy should therefore be masked, particularly on the reverse flanks, by such local advantages as the nature of the country may furnish.

Fig. 70. 9. Echellons of half battalions (*a h*) or less, move by their directing flank (*aa*): Echellons of battalions Fig 71, 73. (*cb*, *ah*) move by their centre serjeants (*s s*): and each echellon of more than one battalion will move in line, or column, by its own regulating battalion, which should always be the one next to the directing flank of the general echellon.

S. 142. *Movements in general.*

Connexion with Battalion Movements.

1. The movements of a brigade, or a considerable line, are derived from those established for a single battalion; and every movement laid down in Part III. can be applied to the evolutions of greater bodies, according to circumstances, to the nature of ground, and to the intentions and objects of a commander: And although each manœuvre given for the battalion, will not be repeated in this part, in its application to the brigade; yet a full discretion can be exercised in the extended application of all such movements, preserving a strict adherence to the principles enjoined, throughout all their details.

Specia attentions

2. The chief objects to be borne in mind are, the exact uniformity of cadence and length of step, throughout all the battalions of a line, and the correct adjustment of distances and points of march and formation by field officers and adjutants.

Inversion of Columns or Lines consisting of several Battalions.

3. Great celerity is given to the movements of extensive bodies, particularly in all changes of front to the rear, by the application of battalion movements on their own ground, so as to invert the order of a brigade or division: This inversion can never be attended with any embarrassment or confusion whatever, provided a battalion is, in itself, never inverted; and a change of front can by this means be accomplished in an extensive line, or in line of contiguous columns, in the same lapse of time as in a single battalion; whereas, when it is performed by the whole line (S. 145), the time of

executi

execution is augmented in a corresponding ratio to the number of battalions of which it is composed.

General Rule. 4. It may therefore be laid down, as a general rule, that great bodies may be inverted, without any liability to confusion, in the order in which their component parts (battalions) may have been originally placed; but, that battalions are never to be inverted in the order in which their companies are arranged by the established formation.

Regulating Body of Movement. 5. All movements, whether in line or in contiguous columns, are regulated by a battalion on either flank, or in the centre. In the advance of contiguous columns at open order, the head of each, (whether they are double or single columns) must conform to the regulating column (*a*); and field-officers must give their special attention that distances for deployment are accurately preserved from the regulating column, whose march will be conducted upon successive points taken up by staff-officers from the proper pivot flank: When contiguous columns are halted, the regulating column will immediately throw out base points for the correction of the heads of columns, (Vide S. 145, No. 3,) and to be in readiness for deployment into line: If the object is merely to preserve a correct order of march, without pressing upon the ordered intervals, a regulating battalion should be named from the centre; but if to make an attack, the regulating battalion should be named from the flank which, being nearest to the enemy, is to give the appui.

Fig. 79.

Advance and Retreat of alternate Bodies. 6. The movements of advance and retreat of alternate bodies are performed in a considerable

line

line either by brigades, battalions, or half battalions, according to the principles laid down, for the battalion in (S. 83 and 84;) the distances between each alternate body being arranged according to circumstances, and the nature of ground: In retiring, one body will protect the retreat of another, either by its number or its position; and when the enemy presses hard, the retreating or rear
Fig. 76—C. line will front and form in the intervals of the first; the whole to be supported by the reserve (*c c*) which, to be in readiness for this object, should lead the retreat: In proportion as the ground
Fig. 77. narrows, different parts of the corps must double; defilés and advantageous posts must be possessed; and by degrees, the respective bodies must diminish their front, and throw themselves into column of march, when it can be done with safety.

7. The retreat, by alternate battalions, or half battalions of a line moving to the rear, while the others remain halted to cover them, and in their turn retire in the same manner, is the quickest mode of refusing a part of the corps to the enemy, and at the same time protecting its movement, so long as it continues to be made nearly parallel to the first position.

Fig. 76—A. 8. If six battalions are in line, the second division, that is, the three even ones (2d, 4th, and 6th,) counting from the right, will go to the right about, retire in line about 200 paces, and then halt, front, having carefully preserved their intervals: The two outward battalions of the retiring ones will each, when it first faces about, form a flank of its outward company (b, b). So soon as the second division begins to retreat, the intervals will be occupied by by light infantry (*a*), and all the battalions of the first one will immediately throw back their wing companies one-eighth of the circle (Vide S. 70, No. 10), and thereby, when necessary, procure a cross fire in the intervals and along the front.

9. When the second division fronts, the first one moves up its flanks, and is ordered to retire through the intervals of the second, (the retreat being covered by the light infantry, which had occupied its intervals,) and to form at an equal distance in the rear: So soon as the first division arrives at the intervals of the second, the latter will begin to fire, as may be ordered: The wing companies of all the second division battalions will place themselves on the flanks, and the light infantry will occupy the intervals so soon as the first division shall pass them; and they will remain so until it is their turn to retire.

Fig. 77. A. 10. When favourable heights or positions present themselves to either of the divisions, they should be for the time occupied by the most contiguous battalions; and any battalions that may happen to possess an advanced height, should throw their wings
Fig. 76—C. back (*e*) in the alignement of their neighbouring battalions, in order to be flanked by them.

11. The retiring division will move by a directing battalion, and any faults in the halt of the line should be corrected before the other division arrives at it. The next retiring division, having the intervals of the previous one to pass, and to move on as a guide, can have no difficulty in its movement or direction.

Refusing a Flank by alternate Bodies. 12. Should the enemy attempt to gain round a flank while retiring in alternate lines, that flank should always be refused, by the retirement of alternate battalions commencing from it: Thus, supposing the right flank turned, Nos. 1, 3, 5 would retreat, leaving Nos. 2, 4, 6 to cover them, and when the latter retreat, they should halt at the interval; so that by continuing the retreat with the right flank always refused, an oblique line can be formed to that flank (*c c*), to repel attack; whereas, if the retreat were led by the other alternate battalions, (*b b*) no such oblique line could be readily formed to the threatened flank: But it may be observed, that if the whole line be formed (C) when a flank is threatened, and that a gradual retreat of the whole force be not called for, such flank should be refused to an enemy by retirement in echellon of battalions, or half battalions, to the extent that may be necessary, (Vide S. 141, No. 6,) which leaves each body free, either to form line (*e e*) to repel attack by a counter one, or to form echellon of squares (*s s s*) should the enemy's flanking force consist of cavalry.

Fig. 76—B

Fig. 76—A.

Fig. 76—C.

Second Lines. 13. The movements of a second line must correspond with those of the first, from which it will always preserve its parallelism and distance. If the first line makes a flank or central change of front, the second must make a change also on such points as will bring it into its relative position.

Fig. 78.

14. When the lines break in columns to the front, the second will generally follow those of the first: When the march is to the flanks, the second line will compose a separate column or columns

When

When the march is to the rear, the second line will lead in columns.

Distance between Lines. 15. Such a distance should be preserved between lines, as may secure the second from any liability to disorder, by the unforeseen retreat of a first line; and, in general, the distance between lines may be equal to the front of two battalions.

16. It must not be supposed that second lines, though so called, are therefore to remain extended; they will generally be formed in column of battalions, or of greater numbers, ready to be moved to any point where their assistance is necessary.

17. The officer commanding the second line must always be properly informed of the nature of the change to be made by the first, that he may readily arrange his corresponding movements.

18. All changes of position of a first line are made according to one of the methods laid down in Part III.

19. When two lines change front to the rear, the second line becomes the first, by each line countermarching upon the ground it occupies.

Fig. 78—A. 20. But when the front is to be changed upon the right or left flank of the first line, the point of distance between the lines will be taken from that flank to the rear; and upon it the corresponding flank of the second line is to form. If the change, for instance, is made on the right flank (*a*), a mounted officer from the second line will mark the distance of that line, in a perpendicular direction to the rear at d 1 or d 2, according as the change is at right angles or oblique to the former front, and the

the second line (*de*) will form upon it (*e* or *o*) by he march of columns, as detailed in S. 158.

Fig. 78—B. 21. But if the change of front is upon a central point (*P*) of the first line (*rs*), the corresponding point will be taken from it, in a perpendicular direction, at (*g*) in the second line; and another point (*Q*) will be taken from the central one (*P*), equi-distant with *Pg*, and accurately perpendicular to the intended line (*bg*). If the change of position is thus at right angles with the former front (*rs*), these two points will be at the extremities of the arc of the quarter circle (*Qg*), of which *PQ* and *Pg*, are the radii: If oblique to it (*co*), they will be at the extremities of the arc (*qg*), composing 1-8th of the circle, more or less, according to the degree of obliquity. The point *Q* for the perpendicular change, and the point (*q*) for the oblique change, will therefore give the corresponding position upon which the central point (*g*) of the second line will be moved by the march of columns, &c., as detailed in S. 159; and this point (*g*), in the second line, will be placed where that line becomes a tangent to the circle drawn from the central point (*P*), provided it is taken up in a true perpendicular direction from the first line.

Passage of Lines through each other.

22. Lines may pass through each other, either in advance or retreat, in any of the methods pointed out in S. 82, particularly when a first line, having suffered in action, retires through a second line, which is brought forward to support it; or when a second line remains posted, the first retires through it; and thus alternately, as occasions may require.

23. In

Crowning a Height in Advance or Retreat.

23. In occupying or crowning a height, by a line, troops should retain, as much as possible, a correct formation; and although the necessity for conforming to the ground, in order to command the approaches to a position, may render it impossible that each body should accurately dress upon the one which precedes it, yet it is essential that every battalion, and, if practicable, every brigade, should be united and square to their respective fronts, in readiness for combined movement to the charge or attack.

Fig. 77.

24. When one line has passed another, it may often, refuse (that is, throw back), a wing, in order to occupy a particular position; to prevent the enemy's design on that wing; or at least to make him take a greater detour to effect it: Or when it is necessary to aligne a corps with a height which is occupied, and from which it may be flanked and protected, as in No. 10.

Fig. 76—C.

Squares.

25. Brigade squares, or any squares consisting of several battalions, can seldom be necessary; for the time unavoidably taken up in the formation is inconsistent with its object, which is generally that of a rapid position of defence against the sudden attack of the enemy's cavalry; and a corps would be placed in a very critical situation if attacked by cavalry during such formation: But battalion squares flanking each other at right angles as explained in S. 79, afford upon all such occasions the most ready and efficient defence; and every battalion commander should carefully keep his corps in readiness for such formations, whether from line or column, to meet any emergency that may offer.

26. When

Fig. 75. 26. When several battalions are drawn up in line, squares may be formed upon each battalion, so as to afford the mutual protection of a flanking fire that will be clear of each other, by the two centre divisions wheeling the eighth of the circle upon the centre (*a*) previous to the formation of squares: They will thus flank each other at right angles (*b*). If when battalions be in echellon, the distance between them, (Vide S. 141, No. 2,) does not happen to be sufficient to bring the flanking fire of squares clear of each other, this object may be attained by the two centre divisions previously wheeling the one sixteenth of the circle, upon the centre towards the advanced echellon, which will make a similar wheel: Upon these occasions the square laid down in S. 89, No. 5, is recommended.

ovements to attack.

Fig. 79.

27. Movements of attack may be made in columns, and the distance from the enemy at which these columns of attack should form in line, must depend upon his position, and the nature of the ground: In general, the line should be formed within 1200 or 1400 paces of a posted enemy, unless the ground particularly favours and covers an attacking force from the fire of his artillery, the enfilade of which is what chiefly prevents bodies in column from approaching nearer; and troops in line will march over that space in 12 or 13 minutes, under the constant protecting fire of their own artillery. Though columns may move with rapidity close to an enemy, yet as they must then form in line, no time would be gained; and their loss would be heavier than when the formation is made at a due distance.

28. When

Reserves and Supports.
Fig. 70—C.
—— 77.

28. When any considerable body of troops is formed for defence, or offensive movement, a portion of it should be placed in reserve (*cc*): And when columns of attack form in line, the reserve should be retained in the formation of column, in readiness to be applied as circumstances of emergency may require: Defensive positions are occupied by lines; and there are situations, according to the nature of the ground, or other circumstances for local consideration, wherein supporting lines (*d*) are an essential formation of defence; but, generally speaking, there is no support so likely to keep an enemy in awe, as that of the formation of columns, provided they can be covered by ground, and protected from the enemy's artillery. The facility and readiness with which columns can be moved and wheeled into any position upon the flanks of a formed line, are calculated to leave an enemy in doubt as to the points where they may be brought to bear against him; whereas lines of support afford a specific position in defence, which invite an arranged attack, and from which it is not always easy to remove troops with rapidity, as occasions may call for a change of formation.

General Remark.

29. The application however of the foregoing principles, and of all the manœuvres and evolutions laid down in this book, must depend upon the circumstances which relatively call for them as occasion may require. For the purposes of elucidation, the effects of formations and movements with reference to the position of the enemy, have been, in some cases, dwelt upon; but all subjects connected with the manner of applying specific formations,

formations, embrace points of consideration much beyond the detail of tactical arrangement; and they can only be governed by all the variety of circumstances, which the nature of the country, and the situation of the enemy, may suggest to the genius of a commander.

EVOLUTIONS OF THE BRIGADE OR LINE.

Plates XI, XII.

The Commands or Caution, given by the Commander of the Brigade, will be distinguished by LARGE ITALICS.

MOVEMENTS FROM LINE.

S. 143. *When a Line formed of several Battalions is required to move to Attack, or to pass a Bridge or Defilé.*

Fig. 79—A.

THE BRIGADE WILL ADVANCE IN DOUBLE COLUMN OF COMPANIES FROM THE CENTRE, (OR OF SUB-DIVISIONS FROM THE CENTRE OF BATTALIONS.)

1. *If to make a movement to attack.*— The advance will be made in so many contiguous double columns from the centre of brigades, or battalions, as may suit the ground, and the circumstances which determine the disposition of attack. The orders for the advance will be circulated: Each commander of brigades will give the annexed caution; and the different commanders of battalions will give the commands laid down in S. 80, observing that if the advance be from the centre of brigades, each battalion composing such double column, will move from its flank (the double column being from

formed upon the right company of one battalion, and the left of another) and its commander must regulate his orders accordingly (Vide S. 86): A regulating column will be named (*a*); and the whole will advance at the Quick March, either upon the signal to that effect, or by observing the movement of the regulating column, to which all the heads of the other columns will conform; points being taken up (dd) and distances preserved, as directed in S. 142, No. 5.

Fig. 20—A.
QUICK MARCH.

All such advances will be protected by artillery and light troops (a, a).

Fig. 79.

2. *If to pass a bridge or defilé.*—The advance will be made from the flanks or centre, according as the defilé may present itself (Vide S. 76, No. 5.) either in one column, or in so many contiguous columns as there are defilés to pass; the passage being covered and protected by artillery and light troops, according to previous dispositions *(a, a, a)*.

HALT
Fig. 79. A.

3. *When columns formed in this manner, either to attack or to pass defilés, shall arrive at the point where line is to be formed,* —they will be halted by signal, or otherwise, and the heads of columns will be dressed on the regulating column (a), which will give base points. The whole will either close to the front of columns and deploy into line (B), or will form line by the echellon march (Vide S. 86, No. 2 and 4,) on the respective heads of columns (A).

4. When

4. When it is intended to advance in several contiguous columns, a greater degree of accuracy in the march may be obtained, by the columns being previously formed at the halt, and then proceeding as already explained.

5. The double columns formed from a considerable line, are susceptible of all the changes of formation laid down in S. 86; it being recollected, that companies hold the relative place in double columns of brigade, that sub-divisions do in battalion: And when a brigade, or line, is formed in contiguous double columns, upon the two centre sub-divisions of each battalion, a rapid change of front, and formation of line to the rear, may be effected by each double column countermarching; the wings at the same time changing places, (Vide S. 86, No. 6,) and the whole deploying into line. This inverts the order of battalions (Vide S. 142, No. 3), but when it is necessary to resume the former front, the original formation is restored by each battalion countermarching on its centre. (Vide S. 91).

6. When a brigade forms double column upon the centre, and when that centre is the right and left companies of two battalions, the interval of six paces will be preserved between the columns; and it will be occupied by a mounted officer (a), who will assist in regulating the square advance of the column.

S. 144. *When a Line formed of several Battalions may have to retire from a flank or centre, or to retreat over a Bridge or Defilé.*

THE BRIGADE WILL CHANGE FRONT UPON THE CENTRE OF BATTALIONS.

1. *If from a flank.*—The caution will be given by the commander of each brigade for a change of front: and each battalion commander will proceed to countermarch his battalion on its own ground, (Vide S. 91.)

S. 91), thus inverting the order of battalions in the line (Vide S. 142, No. 3); and the retreat, that is, an advance to the new front—may be ordered from the flanks or centre of brigades or battalions, as in the last section.

THE BRIGADE WILL RETIRE IN DOUBLE COLUMN FROM THE CENTRE OF——

2. *But a retreat may be effected in double column from the centre without this previous countermarch.*—The commander of the brigade will give the annexed caution; and if the retreat is ordered from the centre of the brigade, the commanders of battalions, on the right and left will WHEEL them BACKWARDS BY COMPANIES on the RIGHT and LEFT, facing that centre: If from the centre of battalions—they will proceed as in (S. 87, No. 2:) And in both cases, the whole will move off at the signal, *Right and left shoulders forward,* to the rear; the heads of columns conforming to the regulating one, and marching upon points taken up as before.

Fig. 24. A.

QUICK MARCH.

HALT.

The whole will halt, upon the signal; the heads of columns will dress upon the regulating column; and if the line is to be formed to the former rear, the columns will change places, by marching through each other; each battalion commander upon the caution, giving the words INWARDS FACE, and proceeding as directed in (S. 87, No. 4.) If the line is to be formed to former front, the columns will not change

THE COLUMNS WILL CHANGE PLACES.

 places,

places, but will countermarch by divisions on their own ground, and then proceed to form line, as already instructed.

If the retreat is for the purpose of retiring through defilés, the columns will be formed in proportion as there may be defilés to pass; and the passage and retreat will be covered by artillery and light troops. (Vide S. 131.)

S. 145. *When a Brigade halted in Line is to change its front to the rear upon the centre.*

THE BRIGADE WILL COUNTER-MARCH UPON THE CENTRE.

1. The two centre companies upon which the countermarch is to be made, will be named; and immediately upon the caution, they will countermarch and change places, as directed in S. 91. The battalions of which these companies form a part, and the remaining battalions on each flank of them, will then receive from their respective commanders, the words THREES RIGHT AND LEFT, RIGHT AND LEFT COUNTERMARCH, according to the flank on which they are placed; and upon the signal to QUICK MARCH, they will countermarch by companies to the right and left, observing that the company on the right of the brigade will circle round to the extent of the depth of a division, to leave room for the other companies to pass each other. The whole will thus form line to the rear as directed in S. 91.

QUICK MARCH.

2. But

2. But the change of front may be effected with greater rapidity, by each battalion countermarching on its own ground, (Vide S. 142, No. 3): And the battalion commanders will proceed accordingly after the caution, as instructed in S. 91; thus inverting the order of battalions in the line.

CHANGES OF POSITION BY MEANS OF THE OPEN COLUMN.

S. 146. *When a Line of several Battalions changes its position upon a fixed Company of any named Battalion by the flank march of Companies in open Column.*

THE LINE WILL FORM OPEN COLUMN UPON (A FLANK OR CENTRAL) COMPANY OF —— BATTALION.

Fig. 80.

1. The commander will name the battalion upon which he intends the change to be made; which in extensive lines will generally be upon the centre; and he will himself see that a company (*e*) of that battalion is wheeled perpendicular to the line he intends to form, whether that line is to be at right angles or oblique, to the old one. The line will then FACE BY THREES, and will proceed exactly as directed in S. 76, No. 7, and in S. 92, 93, and 94, to form open columns upon the pivot flank of the formed company, the adjutants taking up the points of distance (*dd*) between battalions, upon which the leading flanks of their respective battalions are to rest: The whole will form right and left wheel into line.

Fig. 33.

COLUMN WILL FORM LINE.

 2. In

2. In these cases, the adjutants must make themselves acquainted with the total number of files in the battalions on their right and left; for in taking up points where the head companies of their own battalions are to form in the new alignement, they must judge the distance to be occupied by the battalion column to be formed before them.

3. This mode of changing front by means of the flank march of companies in open column, may always be adopted to the extent of a brigade, and always the brigade upon whose
Fig. 80. flank or central battalion, the change is made: But when a line is more extensive than one brigade, it may be expedient that the other battalions should move off, in column of companies *(aa, cc)*, as instructed in S. 86, No. 1, from the right or left flanks, according to which is next to the fixed company (*e*) on which the change is made; and they will successively wheel by companies, or move by the flank march of threes into the new alignement, entering *(d d)* where their head company is to rest, (Vide S. 96;) or where the centre is to rest, (Vide S. 98).

4. When battalions move to a new alignement to the front *(c c)* in this manner they will march off from line, as
Fig. 80. directed in S. 86, No. 1; but when they move to the rear, they will previously wheel backwards by companies (*a a*), and then proceed in column as directed in S. 76, No. 11; and when moving in this manner in column to the front or rear, divisions may be closed to quarter distance to render the march more compact.

S. 147. *When a Line of several Battalions is to change its Position to any distant point, either oblique, or at right angles to the old Line.*

THE LINE WILL CHANGE ITS FRONT IN BATTALION OPEN COLUMNS, TO THE RIGHT (OR LEFT) FLANK. Fig. 81—A.

1. The commander will give the general direction of the new position; and he will signify on what battalion the line may ultimately form; such battalion (*a*) will become the regulating one (which in taking a distant direction will generally be on the flank); and the whole will move as in the last section; except that the battalion (a) upon which the formation is to be made will be moveable until the commander shall fix it in the intended direction; that battalion and the others in the brigade of which it forms a part may move by the flank march of companies, if the ground will permit; the others in column, moving, as before, from the flank nearest to the regulating battalion.

Fig. 81—B.

2. The adjutants will take up their points in the intended line, in the direction which the commander and his staff will indicate, when the company (*c*) upon which the columns are to form is fixed: The whole will proceed as in the last section; and the columns will enter the alignement as instructed in S. 95, 96, 97 and 98, as they may severally apply.

It may here be again observed (with reference to S. 76, No. 6), that in all these movements of battalion columns to form

form in new alignements, the greatest care must be taken by battalion mounted officers that they are kept in constant readiness to form squares (if threatened by cavalry) whether from the flank march of companies in open column, or in direct column of companies. (Vide S. 103, No. 1 and 3.)

S. 148. *When a Column, consisting of several Battalions in Mass, at open or half distance, is required to form a double Column.*

FORM DOUBLE COLUMN FROM THE FRONT.

Fig. 82—A.

1. *If the right is in front.*—The rear half of the column will stand fast; the front half of the column will successively countermarch by companies from the front to the rear, (Vide S. 102, No. 2,) along the right flank (*r r*) of the standing column, and will form column abreast of it (*c r*).

2. *If the left is in front.*—The movement of the front half will be made from the front, in the same manner, along the left flank (*l l*) of the standing column.

The double column may be thus formed for attack, and may deploy or form line by the echellon march, or change front to the rear by the countermarch, and change of places of columns. (Vide S. 86, No. 6.), &c. &c.

Fig. 82—B

3. The formation may take place more rapidly by the column advancing (*a* d), and turning in successive divisions to the rear, until the front division of the rear half (d) shall be in line with the rear division (c) of the front half, when the whole will be halted, and the double column is complete, as above.

S. 149. *When an open Column, composed of several Battalions, shall be required to form a Square or Oblong.*

FORM COLUMN OF GRAND DIVISIONS.

FORM SQUARE ON THE FRONT GRAND DIVISION.

1. The commander of each battalion will proceed upon the caution to form column of grand divisions by the diagonal march, (Vide S. 37 :) The column of grand divisions will then form square upon the front, as laid down in S. 103, No. 1; observing that the leaders of companies shall give the words *Sub-divisions outwards,* when the grand divisions that are to form the flank faces have reached sub-division distance.

2. The square is reduced as pointed out in S. 107, No. 3. The column will open out from the front or rear. The grand divisions will be halted at their respective distances by field officers: The column will then reform column of companies by the diagonal march.

Fig. 47. 3. A double column of companies formed from line at sub-division distance on the two inward flank companies of any two battalions may form square in the same manner.

4. Should it be necessary to move in square, it may be done by FACING INWARDS to the face by which it is intended to move, or in the manner instructed in S. 89 and 90.

QUARTER DISTANCE AND CLOSE COLUMN.

S. 150. *When a Line of contiguous Battalion Columns at quarter or close distance, is required to form a Mass of Columns upon any one Battalion.*

Fig. 83—A.

THE LINE WILL FORM MASS IN REAR OF THE RIGHT BATTALION. THE REMAINING BATTALIONS WILL GO TO THE RIGHT ABOUT.

1. *If in rear of the right battalion.*—The caution will be given, and the commanders of the remaining battalions will give the word RIGHT ABOUT FACE, and they will move QUICK MARCH to the rear upon the signal: The adjutants will be in readiness to take up the distances (*ddd*) of battalions successively at the proper pivot flank from the rear of the formed one (*a*), and when the proper right of each retiring battalion shall reach the exact line of the adjutant, which a field officer (*b*), will accurately mark, the commander will give the word LEFT TURN (or threes left), and when the pivot flank of the battalion shall reach the alignement (*dd*), he will give the word HALT, FRONT, and the whole will cover upon the formed battalion.

THE LINE WILL FORM MASS IN FRONT OF THE RIGHT BATTALION. REMAINING BATTALIONS WILL ADVANCE

Fig. 8—B.

2. *If in front of the right battalion*—The remaining battalions will advance, each commander giving upon the signal, the word QUICK MARCH: The adjutants will take up the distance from the front of the formed battalion (*b*), at which the rear of their respective battalions will rest in the column;

column; and when the rear of each battalion shall get on a line with the adjutant (a), which a field officer will mark (o) as before, the commanders will give the word THREES RIGHT, or RIGHT TURN, and will HALT, FRONT, in the alignement.

The same operation takes place for the formation of mass to the front or rear of the left battalion; the battalions which advance or retire turning to the proper flank when on a line with their adjutants.

Fig. 83—C.

3. *If upon a central battalion.*—The caution will name which battalion: The battalions on the left of it will receive from their respective commanders the word RIGHT ABOUT FACE: The named battalion will stand fast, and the whole will advance QUICK MARCH upon the signal; and will form to the front and rear of the central battalion as in Nos. 1 and 2.

4. When a line of contiguous battalion columns thus form in mass, the battalions need only advance or retire perpendicular to their own front or rear, until clear of the columns on the right or left, when each commander will cut off the right angle (c), by the echellon movement of sections (Vide S. 106, if his column is at quarter distance;) or by the diagonal march. These modes of movement give him a facility for conducting his column to the destined point, without the delay and fatigue of going over unnecessary ground: And this applies more particularly when the mass is to be formed from contiguous columns at open order.

S. 151. *When a Mass of Battalion Columns, at quarter or close distance, wheels by Battalions into a Line of contiguous Columns.*

THE COLUMN WILL WHEEL BY BATTALIONS INTO LINE OF CONTIGUOUS COLUMNS. DOUBLE MARCH.

Fig. 69. A.

1. The caution will be given by the commander; and each commander of a battalion will give the word RIGHT OR LEFT WHEEL (Vide S. 109), and the whole will wheel by battalions at the same moment at the DOUBLE MARCH upon the signal from the commander, dressing to the flank (*b*) to which the wheel is made.

S. 152. *When a Line of contiguous Battalion close Columns is required to change its Front to the right or left Flank. (Vide S. 139, No. 7.)*

Fig. 84.

THE LINE WILL CHANGE FRONT TO THE RIGHT FLANK. ADVANCE IN ECHELLON OF BATTALION COLUMNS FROM THE LEFT.

1. *If to the right flank.*—The caution will be given that the line will advance in echellon of battalions from the left flank (*a*): The commander of the left battalion will then give the word QUICK MARCH; the next battalion will follow when the leading flank of the left one (*a*) shall have gained a space equal to its own front, and six paces (*b*): Each battalion will successively follow in echellon when the one in its front shall have gained this space which (to be accurate) should be marked by a field officer from each following battalion: When the whole are in echellon (*cd*) at this respective distance from each other, the commander

HALT.
THE ECHELLON WILL WHEEL BY BATTALIONS TO THE RIGHT.
FORM LINE ON THE FRONT BATTALION.
QUICK MARCH.

commander will give the command or signal to *Halt*; which should be simultaneously obeyed in order to preserve the distances: The echellon will then wheel by battalions to the right, the commander of each battalion giving the word RIGHT WHEEL; and after the wheel the adjutants will take up their distances from the flank of the battalion (*c*) which becomes the leading one by that wheel; and it should give base points for the others to form upon: The remaining battalions will then form line upon it, being HALTED by their commanders upon their respective adjutants.

The line of contiguous columns will then be in readiness to deploy into line; or to move to the new front; as may be required.

Fig. 69. DE.

2. *If to the left flank.*—The same movement will take place in echellon of battalions from the right; and the battalions will wheel and form to the left (c).

S. 153. *When a Brigade, formed in contiguous Battalion close Columns at close order, is required to change its Front to the Rear.*

THE BRIGADE WILL CHANGE ITS FRONT TO THE REAR ON THE ——— CENTRE BATTALION.
QUICK MARCH.
Fig. 85—A.

1. The caution will be given that the brigade will change its front to the rear on a centre battalion. The commander of the named battalion (*b*) will immediately give the caution that it shall countermarch by sub-divisions (Vide S. 110): And upon the word or signal to *Quick March* from the

the commander of the brigade, the centre battalion will proceed to countermarch at the DOUBLE MARCH, while (supposing the brigade to consist of three battalions, the battalions on its flanks (*c d*) will advance; that on the left (*c*) wheeling on the moveable pivot (LEFT SHOULDERS FORWARD) when it has gained the extent of a sub-division, the battalion on the right (*d*) wheeling RIGHT SHOULDERS FORWARD when advanced, so as to be clear of the left battalion (*c*). The adjutants will have taken up distances to the new front, and base points will run out from each of the flank battalions, and will cover upon the base points which will have been given by the centre battalion on completing the countermarch.

The line of contiguous columns may then move or deploy to the former rear, as may be required.

Fig. 85—B.

2. Should the brigade consist of four battalions, the change of front will be made on the right or left centre battalion: In this case one battalion will wheel round at the depth of a sub-division, as before, along the former front, and the two other battalions will move in echellon from the right, and will wheel round upon the principle given in S. 139, No. 7, when advanced beyond the outer flank of the battalion: Or one battalion (*c*) may go to the RIGHT ABOUT, and countermarch with the rear ranks in front RIGHT OR LEFT SHOUL-

DERS

DERS FORWARD, round the former rear, and HALT FRONT in the alignement, while the two battalions (*bb*) from the other flank proceed round the former front as before.

3. The line of contiguous columns however may, when consisting of any number of battalions, change its front by the countermarch of each battalion on its own centre (Vide S. 142, No. 3,) thus inverting the order of battalions. After this change of front, the line of contiguous columns may deploy; and upon any subsequent occasion the original order is resumed by each battalion countermarching in line upon its own centre. (Vide S. 91.)

4. If the change of front is required, merely with a view to deployment to the former rear, it is of no importance which flank of battalions is in front; and in this case, each battalion column may countermarch by ranks on its own ground (Vide S. 56), instead of by sub-divisions as in S. 110.

DEPLOYMENTS.

S. 154. *When a Mass of Battalion Columns at close or quarter distance, deploys into line of contiguous Battalion Columns.*

Fig. 86—A. *THE COLUMN WILL DEPLOY INTO LINE OF CONTIGUOUS COLUMNS ON THE —— BATTALION.*

1. *If upon the alignement of the battalion on which the column is to deploy.*—The adjutants will be ready to take up the distances for their respective columns in prolongation of that battalion, whether it be a front, central, or rear battalion of the mass. The respective commanders of the remain

ing

THREES RIGHT (or LEFT, or OUTWARDS). QUICK MARCH.

ing battalions will give the word THREES RIGHT (or LEFT) to the proper flank; and the whole will move, upon the signal to deploy, on the named battalion, by the words FRONT or REAR TURN from each commander, exactly as explained in S. 113.

Fig. 86—B.

2. *If upon a rear battalion, but in the alignement of the front battalion.*—The caution will be given accordingly; and the adjutant of that battalion will mark the left flank *(a)* of the front battalion: The whole will then face to the proper flank as before; the named battalion (*b*) will advance to the adjutant, when the front is clear, and the deployment will take place as in S. 113, No. 4.

S. 155. *When a Line of contiguous Battalion Columns at Close Order, deploys into Line.*

Fig. 87—A. *THE COLUMNS WILL DEPLOY INTO LINE UPON ——COMPANY OF —— CENTRE BATTALION. QUICK MARCH.*

1. *If upon a rear or central company of any battalion, and in the alignement of that company.*—The caution will be given, the battalion commanders will face their battalions by THREES to the proper flank, and upon the signal, the whole will move off together, and the named battalion will proceed to deploy to the rear upon a central company *(o o)* in the manner instructed in S. 113; while the battalion *(b)* on the right of the named battalion, will move to the flank, and deploy at the proper interval from the central battalion on its rear company (*r*) (supposing the right of the columns in

in front): The companies in rear of the base (*o o*) *halting* and *fronting* until their front is clear; and those in front of it (*c c*) *turning* to the *rear*, and forming in succession on the alignement: The battalion on the left (*d*) will form from the right to the rear, by the right company (*d*) receiving the word *Rear turn, Halt,* when it shall be at the proper distance (d) from the left of the battalion on which it is to deploy: The right company will be halted in this manner by a mounted officer, who will be able to judge the proper distance from the left of the preceding battalion: Each company of the right wing will form to the rear in the same manner, and when their front is clear, will successively move into the alignement with the base (*o o*); and the companies in rear of that base will *front turn* in succession into the alignement.

Fig 87—B. 2. *If upon a front company of any one battalion; or upon a rear or central company in alignement with the front.* The caution will be given by the commander, as before, and the deployment upon a front base will be performed exactly as laid down in S. 113, No. 4; the named battalion (*a*) deploying as if it were a single battalion; and each other battalion forming upon the front or rear companies, according as they are on the right or left of the named battalion; each company *halting* and *fronting* until its front is clear, or when the front is clear, *front turning* into the alignement.

In

In these deployments the points and distances will be taken up by field officers and adjutants, as explained in S. 136, No. 7.

S. 156. *When a Mass of Battalion Columns, at close or quarter distance, is required to deploy into Line, facing to the rear.*

Fig. 88.

The mass will deploy into contiguous battalion columns in inverted order, that is, to the reverse flank (*b*), (supposing the right in front): Each battalion column will then countermarch upon its centre, (Vide S. 91); or by companies or ranks (Vide S. 56,) which will rectify the inversion, and at the same time change front to the rear; the latter countermarch bringing the left of columns to the front (*c*). The whole will then deploy into line upon any company of any battalion.

S. 157. *When a Mass of Battalion Columns at close or quarter distance, shall be required to deploy into Line, to any distant position on detached Adjutants.*

Fig. 89.

1. The mass will disengage to the proper flank (*b*) either by deploying into a line of contiguous battalion columns, or by breaking into an echellon of battalion columns (*b c*). The general direction of the new line will then be given (*aa*), and the regulating battalion (a flank one) will be named (*c*), according to the flank to which it

it is intended to incline. The adjutants(*aau*), will take up the new alignement where the leading division of each battalion is to rest, in a contiguous line of columns, either at close or open order; and each battalion commander will conduct his battalion to his adjutant; accommodating his march to the nature of the ground, and diverging as may be necessary, from time to time, by the wheel on the moveable pivot, the diagonal march, or the echellon march of sections, into the proper direction: If the battalions form in the new alignement in close order, the line is formed by a general deployment (Vide S. 155): If at open order(*a a*), each battalion will deploy on its own ground.

2. In taking up distances in a contiguous line of columns at open order (*a a*), it is essential to observe, that Fig. 89. if the right of the battalion columns be in front, and that the general alignement is to be taken up from the *right*, each adjutant must judge the distance which the battalion on his right is to occupy: If the left of the battalion columns be in front, each adjutant will judge the distance which his own battalion is to occupy: But in taking up the general alignement from the *left*, if the right of the battalion columns be in front, each adjutant must judge the distance for his own battalion; but if the left of the columns be in front, he must judge the distance to be occupied by the battalion on his left.

3. When battalion double columns assemble in a Fig. 79—B. contiguous line for deployment, the adjutants will take up the distances (*d d*) where the centre (that is,

the head of each column) is to rest; and each adjutant must take a distance equal to the extent of one of the wings of his own battalion, and the wing of the next battalion: If double columns assemble in close order, they must open out to those distances before they can deploy.

SECOND LINES.

S. 158. *When two Lines change position upon a flank of the First Line.* (*Vide* S. 142, No. 20.)

Fig. 90.

1. The first line will proceed to change its front to the given direction (*a c*) either by the echellon march of companies, or by the open column, as directed in S. 92. A point will be taken (d) at a distance equal to the space which separates the two lines, and exactly perpendicular to the intended line; and this point marks where the right (*d*) of the second line is to form.

Fig. 90.—A.

2. Base points will be given (*o o*) to the right of the point (d), and the outer flank (*c*) of the first line being ascertained, a corresponding point (*e*) will be taken for the outer flank of the second line by an officer pacing in a perpendicular direction from (*c*) the same distance which was taken for the other flank (d): A point will then be taken up in prolongation of *e*. d. *o*., which intersects the original second line at *S*; and that part which is to the right of the point *S*, will be wheeled into open column, left in front, while the line to the left of the point

point *S*, will move off from the left of companies by threes, and form open column on the alignement, *(S e.)* Vide S. 92. This part of the column will then go to the right about, and will prolong the alignement to the required extent; while the portion of the old line *(S d)* will follow in column, wheeling into the new direction at the point *S*,: And when the rear of the column shall arrive at d, the whole will HALT, the part which had faced about will FRONT, and the whole will RIGHT AND LEFT WHEEL INTO LINE.

Fig. 90—B

3. Or, the second line may form columns (*m*) upon a flank of each battalion (on the left flank when the change is to the right, and upon the right flank when the change is to the left); the contiguous columns will then take the necessary ground to the left, and will halt and front, when the right column has its right flank (m) opposite the point (d) on which it is to form in the new line: The different columns will then move to the alignement (d *e*) and will form at open order upon their respective adjutants, who will take up the distances as directed in S. 157; each commander guiding his battalion column to his adjutant, by the combined application of the wheel on the moveable pivot, the diagonal march, &c., as may be necessary.

S. 159. *When two Lines change position upon the central point of the first Line.* (*Vide S.* 142, *No.* 21.)

Fig. 91.

1. *P* is the point in the first line on which the change is made: A mounted officer will mark the corresponding point (*g*) in the second line: Another point is taken at *Q*, in a perpendicular direction to the intended line: Points are taken up (*o*) to mark the alignement with the outer flank of the intended line, as in the last section.

2. The first line then changes its front or position upon the centre *P*, by the echellon march, or the open column, as directed in S. 92. A point (*S*) will be taken up in prolongation of *o Q*, and that portion of the second line which is to the right of that point (*S b*) will be thrown into open column, left in front; while the other portion of it (*S h*) will move off from the left of companies by threes, and will form open column in the alignement (*S o*) as in the last section; it will then be faced about to prolong the alignement, while the column (*S b*) will successively wheel at *S* into the new direction, and the whole will be halted when the central division (*g*) shall reach the central point *Q*; the rear companies of the column will move, if necessary, into the alignement, and the whole will form line, as in the last Section.

3. Or

3. Or the second line may change its position as in the last Section, No. 3, by forming columns on the flanks of battalions, and proceeding as therein directed: or the line may break into open column, close to the front of battalions, and taking up the alignement (*S o*) in a similar manner, by the wheel of columns on the moveable pivot and the diagonal march into contiguous positions at open order.

These changes of position being to the right flank, it is only necessary to observe that the same instructions apply to changes of the left flank.

END OF PART V.

INSPECTION

INSPECTION OR REVIEW.

The battalion formed in line at open order, will await the approach of the general. He is to be received with the compliments due to his rank, as set forth in the regulation of military honours. The colonel and lieutenant-colonel on thi occasion, are on foot, at the head of the colours, and at all other times they are to remain on horseback.

A camp colour is to be originally placed 80 or 100 paces in front of the centre of the battalion, where the general is supposed to take his station; but although he may choose to quit that position, still the colour is to be considered as the point to work upon, and to which all movements and formations are relative.

Receiving the General.

PRESENT ARMS. — When the reviewing general presents himself before the centre, and is 50 or 60 paces distant, he will be received with a general salute. The men present arms, and the officers salute, so as to drop their swords with the last motion of presented arms; the music will play, and all the drums will beat. The colours only salute such persons as from their rank, and by regulation, are entitled to that honour.

SHOULDER ARMS. — The men shoulder, and the officers recover their swords with the last motion.

The general will then go towards the right, the whole remaining perfectly steady, without paying any further compliment while he passes along the front of the battalion, and proceeds

ceeds round the left flank and along the rear. When the general is going round the battalion, the music and drums may play and beat; they will cease when he has returned to the right flank of the battalion.

REAR RANK TAKE CLOSE ORDER, MARCH. — While the general is proceeding to place himself in the front, this command will be given, and the colonel and lieutenant-colonel will then mount on horseback, in the rear of the centre.

Marching past in Slow Time.

COMPANIES ON YOUR LEFT BACKWARD WHEEL. QUICK MARCH, *Halt, Dress,* MARCH. — The battalion will break into column of companies, the right in front, and the column will be put in motion, pioneers, music and drums having been previously ordered to the head of it. Points will be fixed by the adjutant for the several wheelings of the divisions; so that their right flanks, in marching past, shall be only four paces distant from the camp colour, where it is supposed the general places himself to receive the salute.

Right Shoulders Forward. — The several companies wheel successively at the first and second angles of the ground on the moveable pivot; and the wheel at the latter angle will bring them on the line on which they are to pass the general. Each leader of a company, when it has advanced six paces from the wheeling point, changes quickly by the rear to the right flank of his company, and as soon as he has placed himself on that flank, he

By the Right. — will order *By the Right*, upon which the men will touch to the right, keeping their eyes direct to their front.

The

Rear Rank take Open Order. The leading company, and each other successively, as it arrives within thirty paces of the general, opens its ranks, at which time the leaders of companies move to the front, and are replaced on the right flanks by their serjeants.

In marching past the reviewing general, the colonel is to be in front of the grenadier company, with the major a little behind him, on his left: The music, drummers, and fifers, are six paces before the colonel, and the pioneers are in two ranks, six paces before the music, having a corporal at their head to lead them.

The lieutenant-colonel is to be in the rear; but in the absence of the colonel, the lieutenant-colonel will supply his place. The second major is in the rear, behind, and on the left of the lieutenant-colonel; and the adjutant is behind and on the left of the major.

The colours, carried by the two *senior* ensigns, are three paces behind the fourth battalion company, covered by their serjeants. Staff officers do not march past.

In marching past at open ranks, the serjeant who is on the right flank of the company is responsible for the proper wheeling distance being kept from the front rank of the company preceding him. The leading officer must invariably preserve his distance of three paces before the right of the company, and not derange its march; the rank of officers dress to him, their eyes are turned a little to the right, and they divide the ground, in order to cover the front of the company: If there is only one officer with the company, he is towards the right of it. Supernumerary serjeants are three paces in the rear of their several divisions.

The music will begin to play just after the leading company has made the second wheel; they will draw up opposite

to

to the general, and they will continue to play until the rear of the column shall have passed him.

The officers when they arrive at their proper distance from the general, will salute successively by companies, when within six paces of him, and recover their swords when ten paces past him, without in the least altering the rate of march, or impeding the front rank of companies. The commanding officer, when he has saluted at the head of the battalion, places himself near the general, and remains there till the rear has marched past.

Rear Rank take Close Order. By the Left.	The officers commanding companies will each successively, when he has passed the general by twenty paces, close his rear rank, and will move by the rear to the left flank, giving the word *By the Left*, upon which the men will lightly touch to that flank: Each individual of the company will at the same time, resume the post which he held when the column was first put in motion.
Right Shoulders Forward. Forward.	The several companies wheel successively at the point which will be fixed opposite the ground where the left of the battalion stood.
HALT. SLOPE ARMS.	When the leading company is near to where the left of the battalion stood, the whole HALT, arms are sloped, and the Quick March will instantly commence.

Marching past in Quick Time.

QUICK MARCH.	The whole march off in quick time. No music.

The

Right Shoulders Forward. Forward. — The column makes three several wheels on the moveable pivot, *viz.*, at the point where the left of the battalion first stood; at the point where the first wheel was made; and at the point where the second wheel was made, which places it on the line of passing the general.

CARRY ARMS. — Before the leading company has made the last named wheel, arms are carried. When it has completed that wheel, the music begin to play.

The leading officer of each company will shift to its right by its rear, in the manner already instructed, six paces after the wheel, which brings him on the line with the general; and when he has passed the general twenty paces, he will resume his proper pivot flank. The supernumerary officers and serjeants march in a rank in rear of the companies, at one pace from the rear rank, and officers' swords must be carried against the right shoulder, steadily.

The colonel, lieutenant-colonel, major, and adjutant, are in the same places as in marching past in slow time; as also music, drummers, &c., which will play drawn up as already directed.

By the Left. — The several companies, twenty paces after passing, will successively touch to the left (the proper pivot flank) and the officers will shift to that flank while giving the word.

Right Shoulders Forward. Forward. — The companies successively wheel when opposite to the ground where the left of the battalion stood: And will again wheel at the next angle, where the left actually

stood

stood, which will bring the column upon the original alignement. The leader of the front company will march upon a point, and each officer will cover in column and preserve distance, in order to be ready for the halt and formation of line.

Forming in Line.

HALT,
LEFT WHEEL INTO LINE.
QUICK MARCH.
Halt, Dress.

The column will proceed on the alignement, until arrived at the point where its head, or right, is to be placed. It will receive the word HALT; pivots are instantly corrected, if necessary; it will then be wheeled into line, and the pioneers and music will go to their posts behind the centre.

When the line has formed, the commanding officer will give a caution that the manual and platoon exercise will be performed, and he will then go to the rear of the battalion. The major advances to the front of the battalion, OPENS RANKS; UNFIXES BAYONETS; SHOULDERS ARMS; makes the officers and colours TAKE THEIR POST OF EXERCISE in the rear, by facing to the right; MARCHING through the several intervals occupied by the serjeants; and when three paces beyond the rear rank, they halt, and then receive the word FRONT: The commanding officer, lieutenant-colonel, adjutant, pioneers, music, supernumerary serjeants, drummers, fifers, are at their posts in the rear, as when the battalion is formed in close order.

The major proceeds with the manual as directed by regulation. The serjeants, who preserve in the front rank the places of the platoon officers, remain there steady during the whole of of the manual, except that they charge their pikes at the same time as the bayonets.

When the manual has been performed, the major will proceed with the platoon exercise as detailed in regulation, either in slow or quick time as may be required.

When the manual and platoon exercises have been finished, the major will go to his post, and the commanding officer of the battalion will unfix bayonets, and prime and load with cartridge, and will then commence such of the movements laid down in the preceding sections, as may be deemed applicable to the nature of the ground, or as may be ordered at the discretion of the general, or the commanding officer, if left to his own selection; observing always, that such of the specified movements as may be practised, are strictly adhered to in all their details; and that not less than ten or twelve of them are performed: Every commanding officer must consider himself ready and liable to be called upon for the immediate performance of any of the movements laid down in this book.

When two or more battalions are inspected or exercised together, they will be formed in one line, with the ordered intervals, and will perform the same movements that are laid down for the single battalion, observing the additional directions that are given for such movements when applied to the brigade. Upon occasions when the line may exceed two battalions, the reviewing general may, at his option, dispense with marching past in slow and quick time in open column, in order to save time, and to preserve the troops fresh for the subsequent movements.

RIGHT SHOULDERS FORWARD. FORWARD. QUICK.

In such cases the line may march past in quick time only, in battalion columns at quarter distance; each battalion wheeling at the respective angles, as instructed in S. 109, No. 4.

To

To prepare for marching past in this manner, the line will form contiguous columns of battalions at quarter distance; and will wheel into mass of columns, and close to the distance of twenty-five paces, between each battalion, to leave room for the field officers, music, and pioneers: Commanding officers of battalions will be careful to preserve this distance in marching past; and corrections may easily be made at each wheeling angle, provided the leading battalion will continue to proceed a sufficient number of paces at the double march after the word FORWARD, (when the wheel has been completed) in order to prevent the other battalions from crowding upon it. It will generally be found, that from fifteen to twenty paces, between the words FORWARD and QUICK, will be sufficient for the leading battalion: The commanding officers of the following battalions will regulate the interval of time between these two commands, as they may find necessary for the correction of distance.

Upon these occasions, mounted officers will salute. All other officers will march at their respective posts in column; but the leaders of companies will change their flank after the wheel which brings them on the alignement where the general is placed, in the same manner as laid down for marching past in open column in quick time.

When the mass of columns which has thus marched past shall reach the original position opposite the general, the major-general, or brigadier, will halt the leading battalion, and the others will close to the front to ten paces interval between each battalion: He will then wheel the mass into contiguous columns, and will UNFIX BAYONETS—PRIME AND LOAD WITH CARTRIDGE, and proceed with his intended movements.

In order to be perfect when formed in brigade, every battalion should be well practised when single, in marching past in columns at quarter distance.

FINIS.

INDEX.

INDEX.

A.

B.

C.

E.

F.

G.

H.

I.

K.

L.

M.

N.

O.

P.

By Authority.

A Company formed Two, Three, and Four Deep, A.B. and C.

The same on the reduced scale.

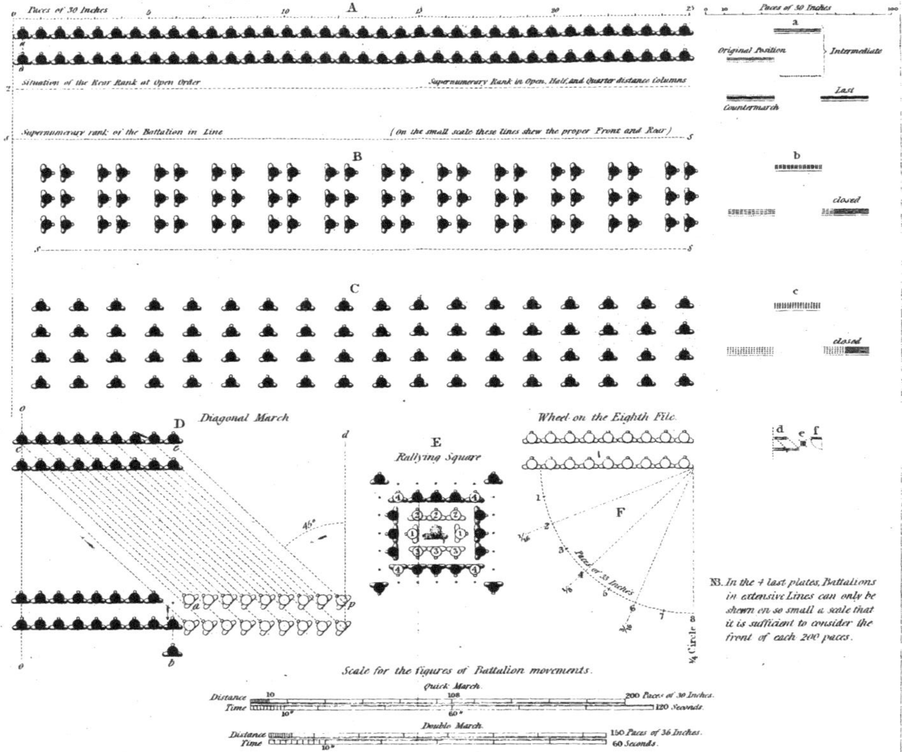

The dotted lines and arrows expreſs the direction of movements.
The round dots, mark the position of leaders of divisions & intermediate covering points.
Square dots expreſs points of March or formation.

Capital letters as A.B. denote different movements in the same figure.
Where any letter of reference &c. as L.G. a.b. 1.2. is repeated thus L.G. a.b. 1.2.
the latter character marks a subsequent position of the same division.

Capt. Mitchell del.

Engraved on Steel for the Adjutant General, 1824.

W. Lowry sculp.

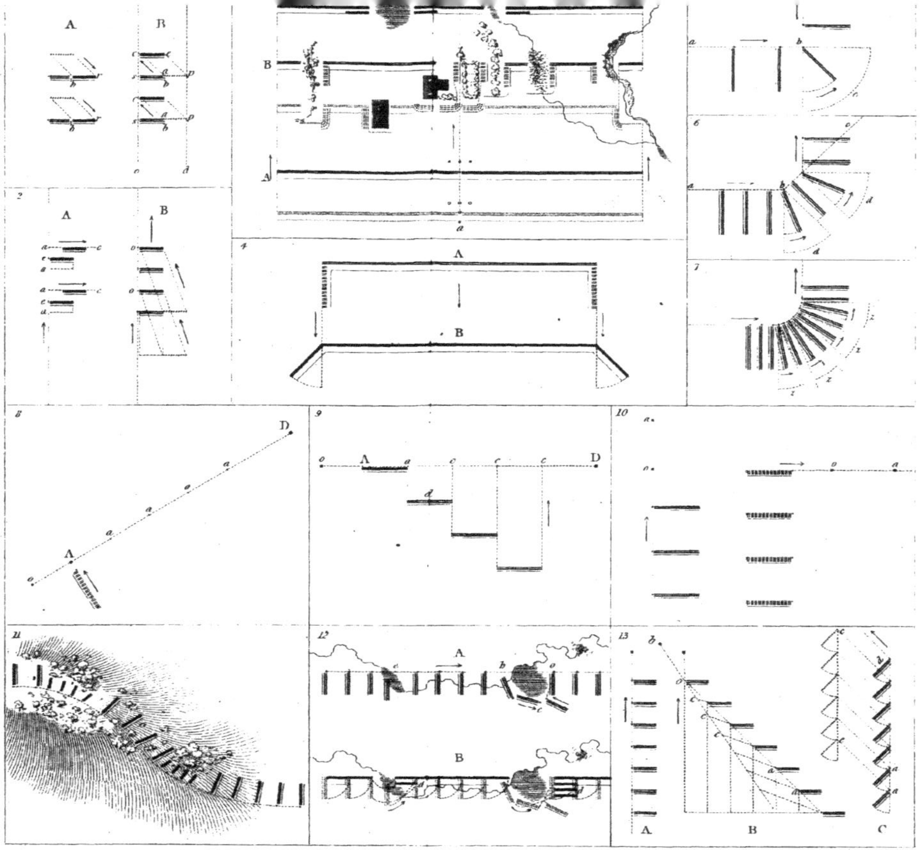

Engraved on Steel for the Adjutant General, 1824.

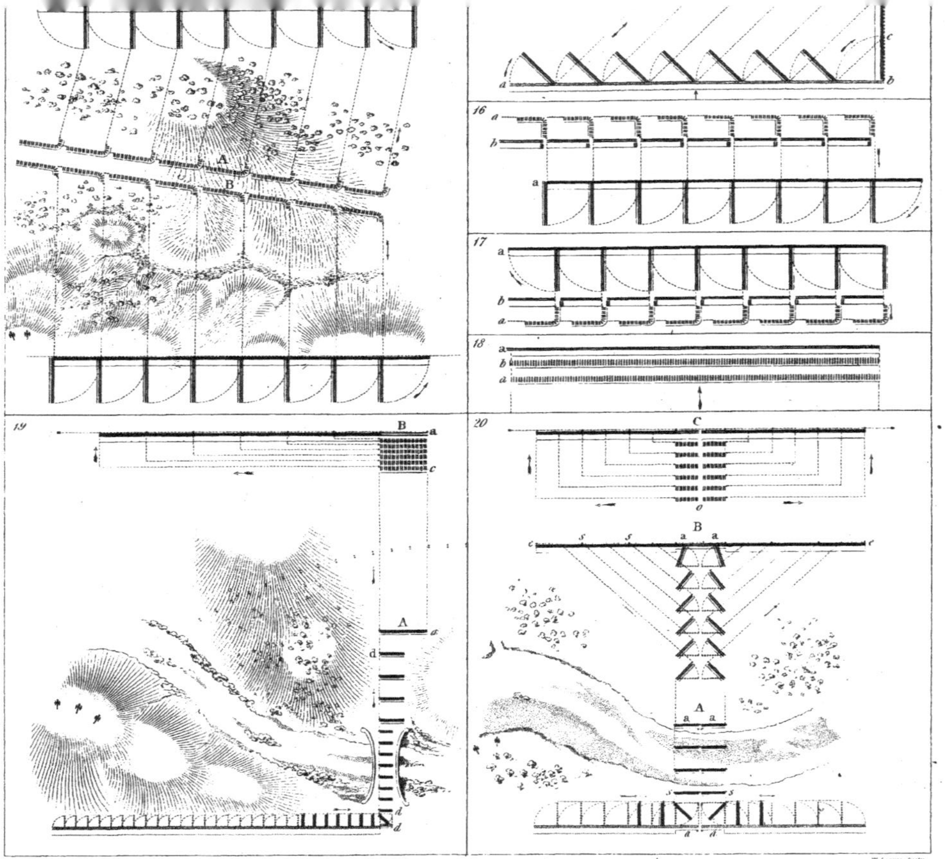

Engraved on Steel for the Adjutant General, 1824.

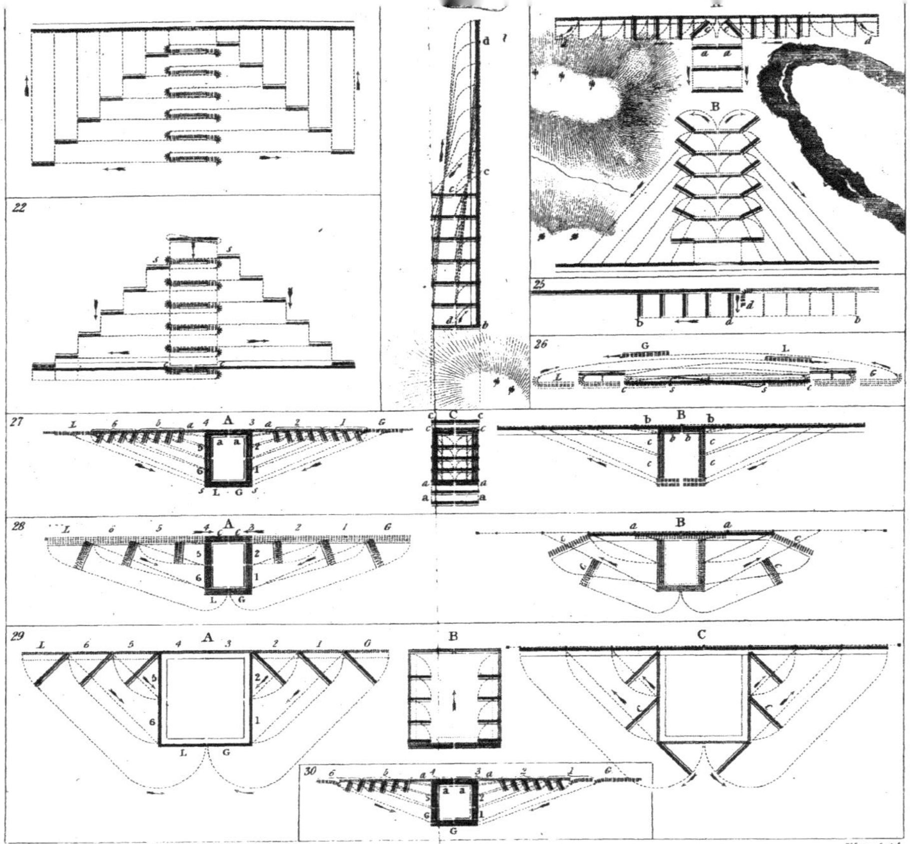

Engraved on Steel for the Adjutant General, 1824.

W. Lowry sculp.

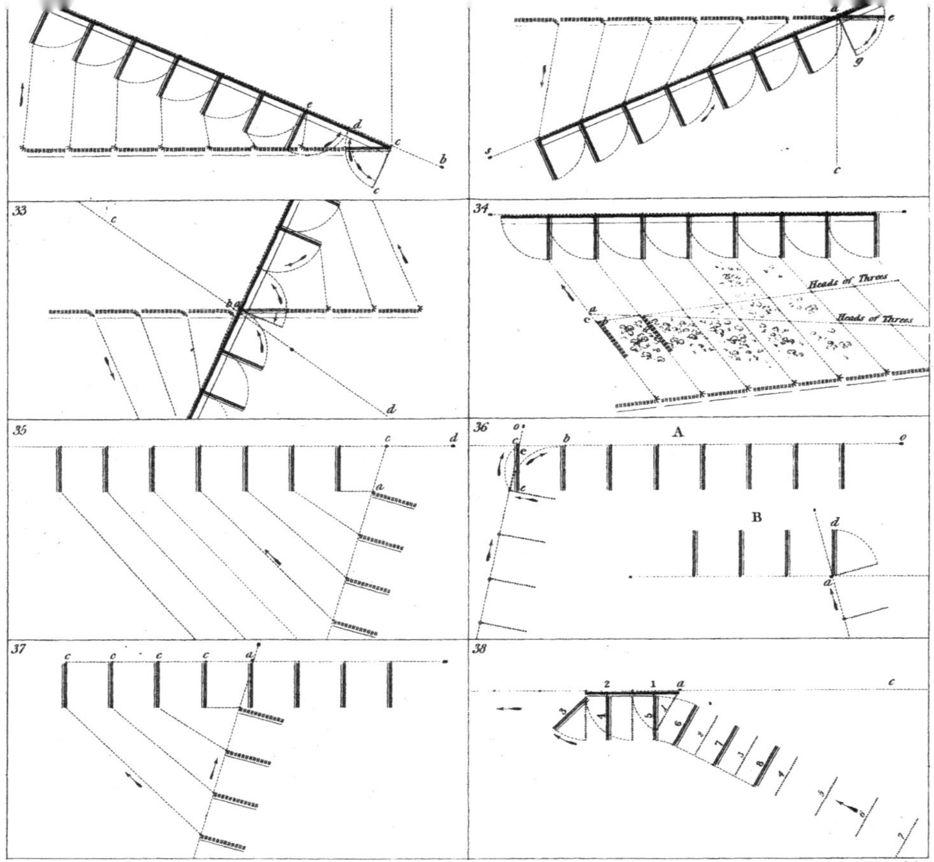

W. Lowry sculp.

Engraved on steel for the Adjutant General 1824.

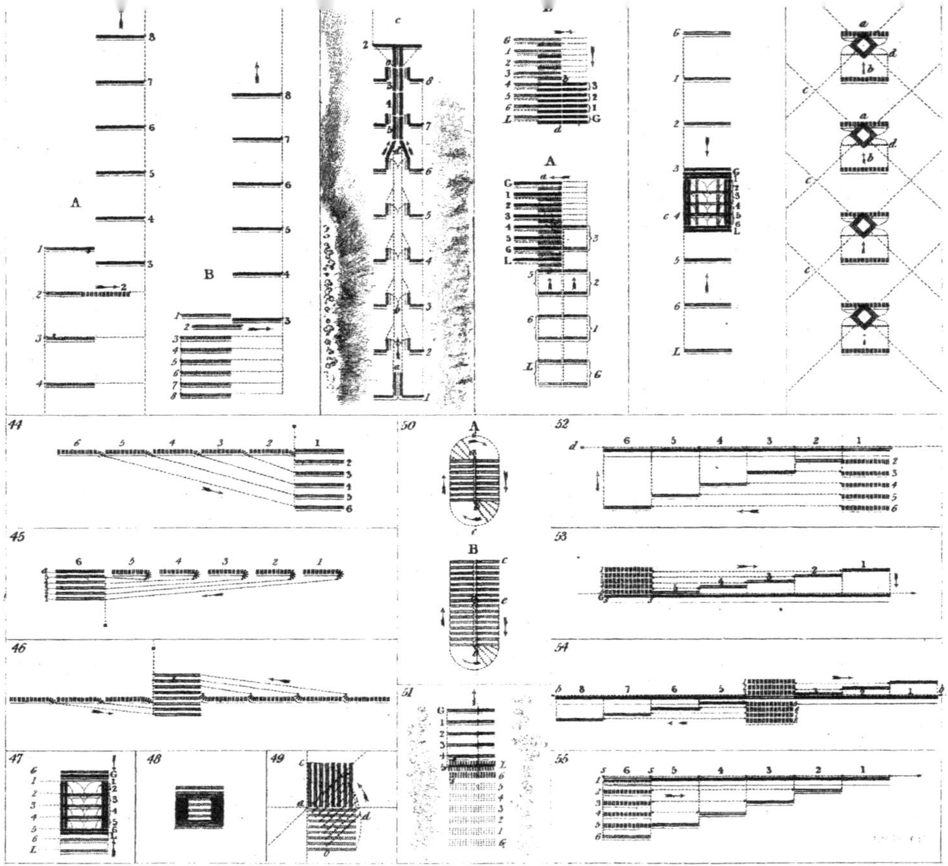

Engraved on Steel for the Adjutant General, 1824.

W. Lowry sculp.

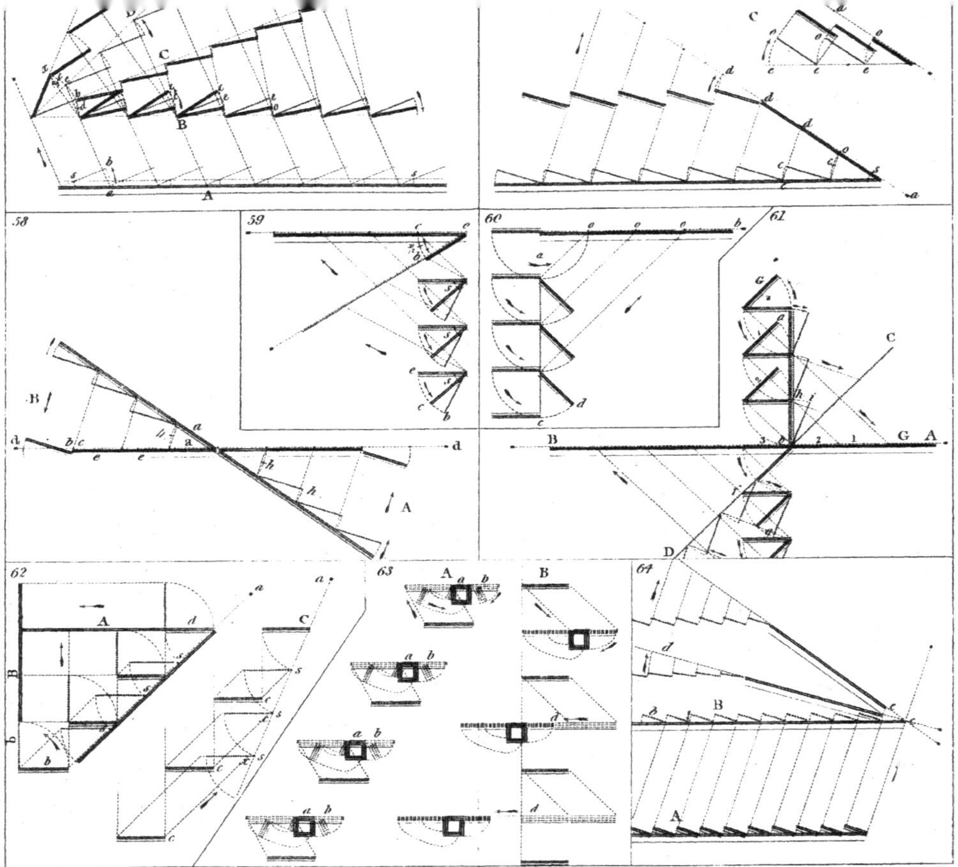

Engraved on Steel for the Adjutant General 1824.

W. Lowry sculp.

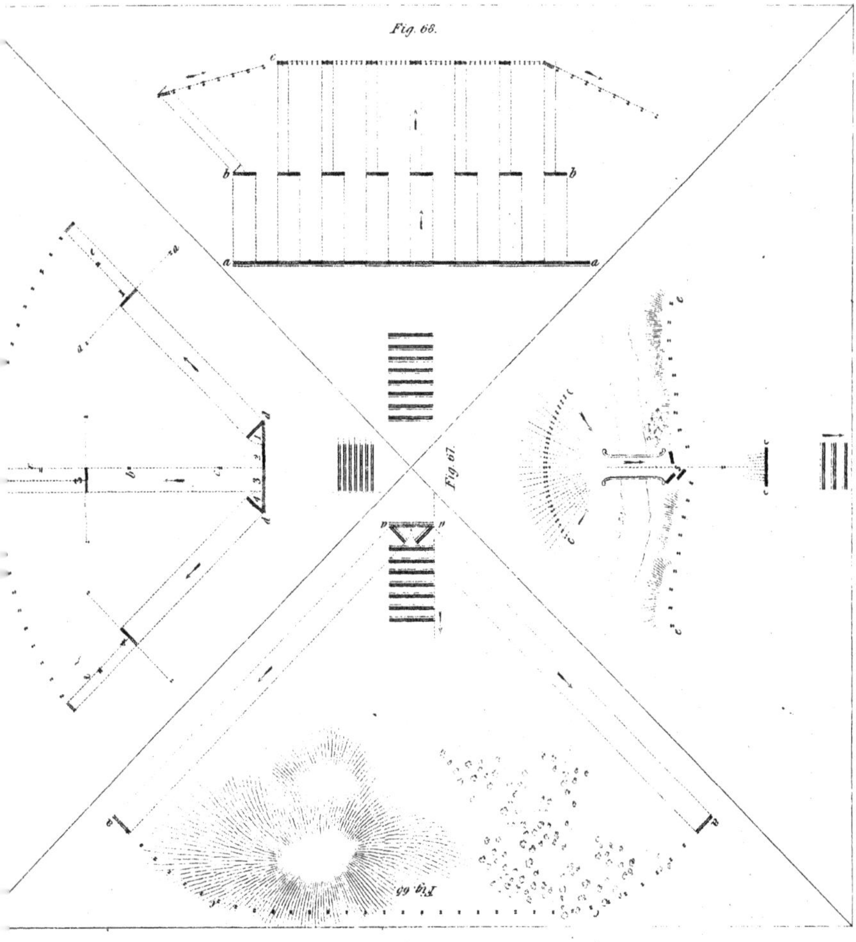

Engraved on Steel for the Adjutant General. 1824.

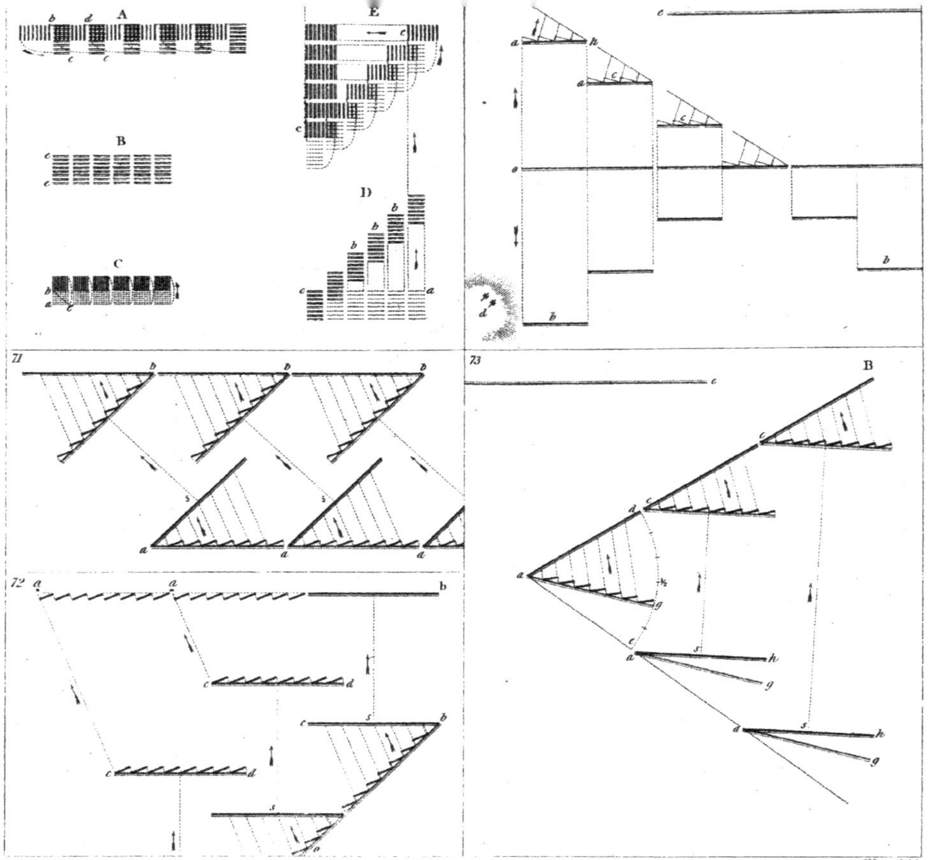

Engraved on Steel for the Adjutant General, 1824.

W. Lowry sculp.

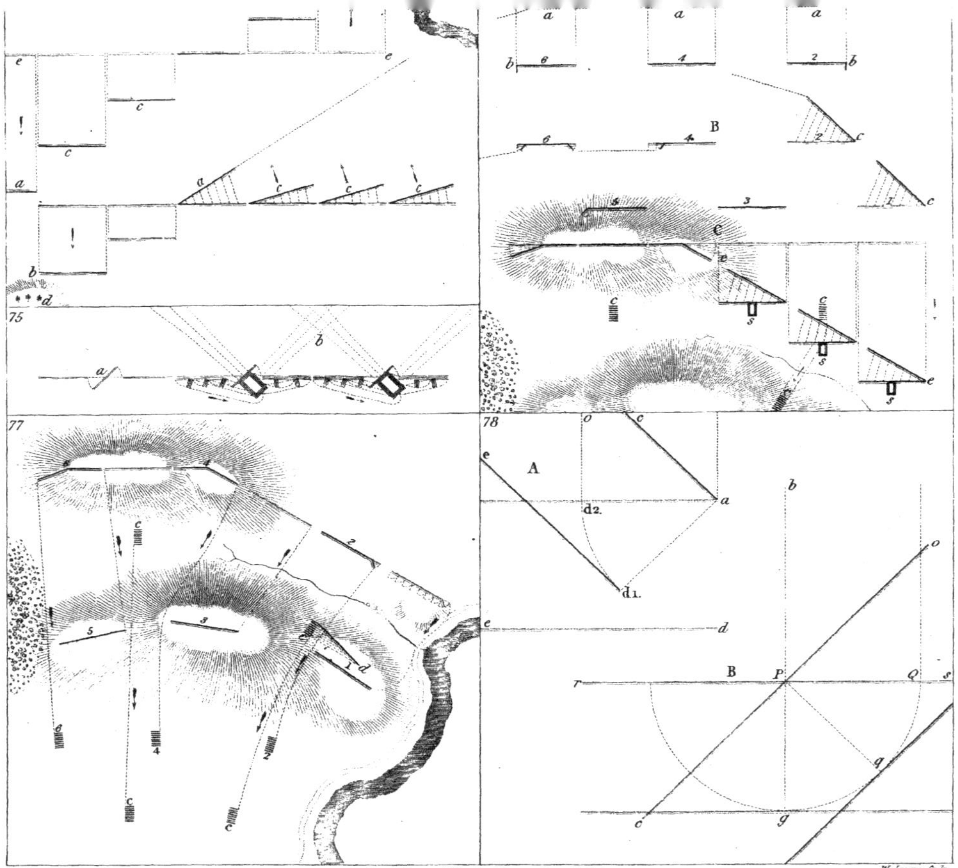

Engraved on steel for the Adjutant General. 1824.

W. Lowry. sculp.

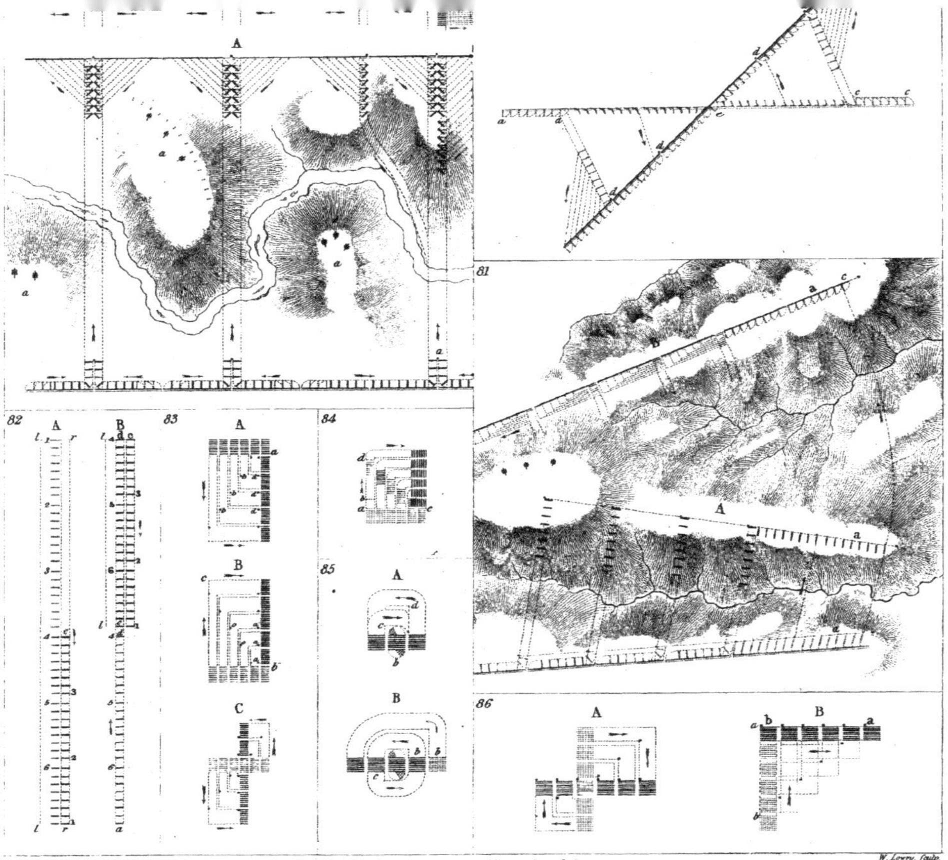

Engraved on steel for the Adjutant General. 1824.

W. Lowry, sculp.

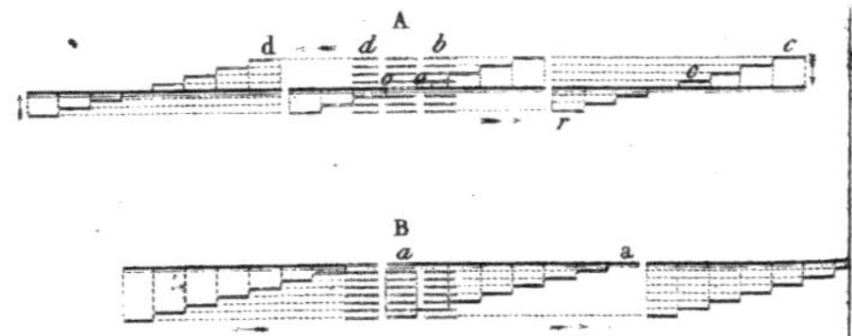

88

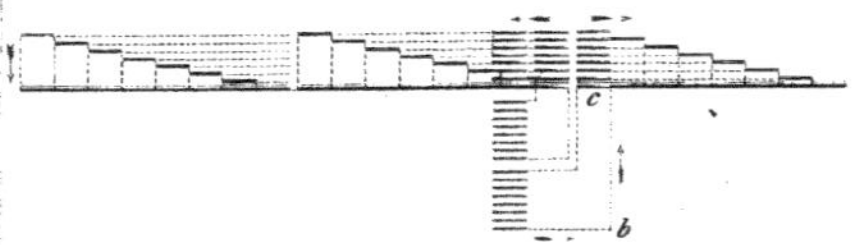

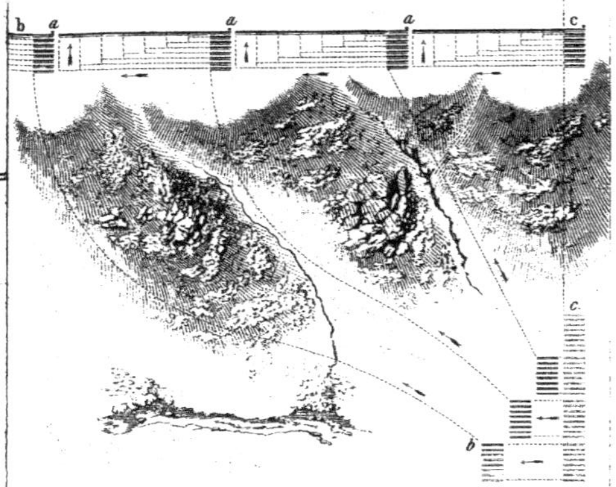

90

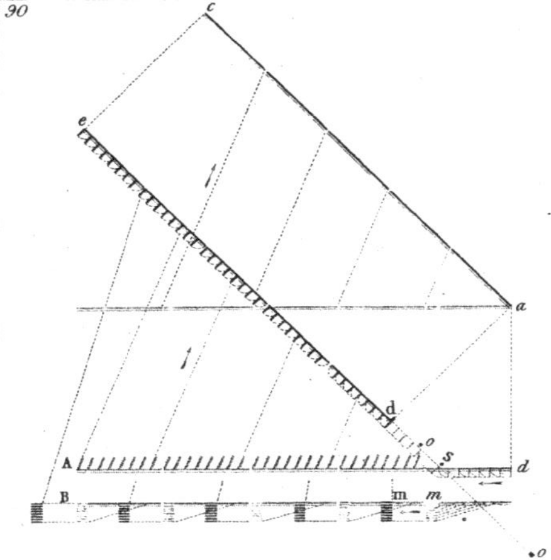

91

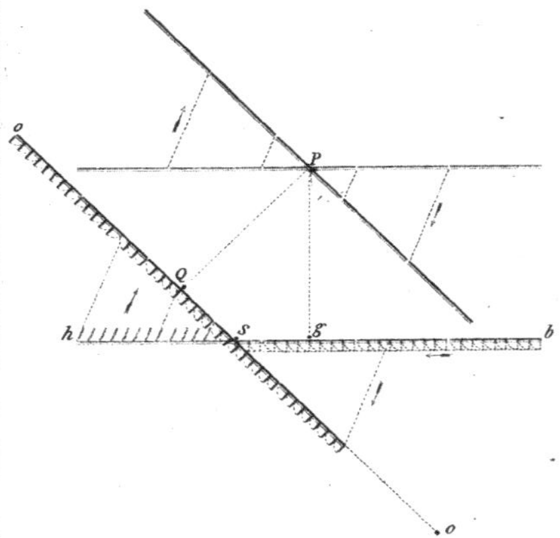

Engraved on steel for the Adjutant General 1824.

W. Lowry sculp.

www.ingramcontent.com/pod-product-compliance
Ingram Content Group UK Ltd.
Pitfield, Milton Keynes, MK11 3LW, UK
UKHW041847031026
15587UKWH00005B/1072/J

9 781847 342317